Maggie Black writes on international issues
for UNICEF, WaterAid and the Global Water
Partnership, among others. Her recent books include
Water: A Matter of Life and Health (with Rupert Talbot),
Water Life Force, *The No-Nonsense Guide to Water*,
and *The Last Taboo: Opening the Door
on the Global Sanitation Crisis* (with Ben Fawcett).

Jannet King was coauthor, with Robin Clarke, of
the first edition of the atlas. She has spent many years
researching and editing political, social, and environmental
atlases, including *The State of the World Atlas*,
the award-winning *Atlas of Food* and *Atlas of Health*,
The Atlas of Endangered Species, and the World Bank series
of *miniAtlases* on development, the environment,
and human security.

In the same series:

"Unique and uniquely beautiful. . . . A single map here tells us more about the world today than a dozen abstracts or scholarly tomes." *Los Angeles Times*

"A striking new approach to cartography. . . . No one wishing to keep a grip on the reality of the world should be without these books." *International Herald Tribune*

THE ATLAS OF
WATER

Mapping the World's Most Critical Resource

Second Edition

Maggie Black and Jannet King

Foreword by Margaret Catley-Carlson

UNIVERSITY OF CALIFORNIA PRESS

Berkeley   Los Angeles

University of California Press, one of the most distinguished university presses in the United States, enriches lives around the world by advancing scholarship in the humanities, social sciences, and natural sciences. Its activities are supported by the UC Press Foundation and by philanthropic contributions from individuals and institutions. For more information, visit www.ucpress.edu.

University of California Press
Berkeley and Los Angeles, California

Cataloging-in-publication data for this title
is on file with the Library of Congress.

ISBN: 978-0-520-25934-8 (pbk. : alk. paper)

Produced for University of California Press by
Myriad Editions
Brighton, UK
www.MyriadEditions.com

Edited and co-ordinated by Jannet King and Candida Lacey
Designed by Isabelle Lewis and Corinne Pearlman
Maps and graphics created by Isabelle Lewis

Printed on paper produced from sustainable sources.
Printed and bound in Hong Kong through Lion Production
under the supervision of Bob Cassels, The Hanway Press, London.

15   14   13   12   11   10   09
10   9   8   7   6   5   4   3   2   1

Contents

PART 3 Water for Living 44

PART 4 Water for Economic Production 60

Foreword

I asked two friends, one Mexican, one West Coast USA: "If I gave you all the money you need to fix the drying Colorado River what would you do?" There was a long silence – a very long silence – then both spoke at once. "It really isn't about money. We need to change the way people think about water. We don't have the public climate of thought to make the right decisions."

How often do we really *think* about this essential, ever-present marvel called water? We bathe in it, drink it – even eat it, given the huge amount needed to grow food. We transport goods by water, we take vacations to be near water, and we fish in rivers and lakes. Industries could not function, energy be generated, nor books or clothing created without water. We send rockets into space to find water on nearby planets. Poets extol its beauty and worshippers purify themselves with it before prayer. Despite all of this, we behave as if water can manage itself.

Surely we should take better care of it? As you will learn in this Atlas, there is a world water crisis today, a crisis of underground water levels declining, serious pollution, lakes and water sources disappearing. As many as 70 rivers are so over-used that they are *closing*, which means that we cannot afford to take more from them. While around 1 billion people have no consistent access to drinking water, 2.5 billion lack sanitation – a threat to their health and dignity and often to the water around them.

It takes about a litre of water to produce one calorie of food. As prosperity increases, so does demand for high-protein, water-intensive food. And it takes up to nine times that much water to produce the same unit of energy, whether by pumping oil, growing bio-fuels, or in the other ways we produce power for heat, light, transport and production. As populations increase, all these demands will skyrocket.

The Earth today has the same amount of water that existed for the dinosaurs. Where that water lies and how we treat the water close to us is the issue at stake. Around half a billion people in 29 countries already face water shortage. The combination of population growth, pollution, and climate change means that by 2025, fully one-third of the expected world population will be in regions facing water scarcity – and paradoxically, at higher risk of water disasters: floods and severe storms.

Where there is strong competition over water resources, two outcomes are predictable: the environment suffers, and poor people have more trouble gaining access to the water they need for drinking and living. There is also a risk of dispute, even conflict, in one of the 260 river basins shared between countries. Development itself becomes more difficult as water-short areas fail to attract investment.

We do have responses for these problems. A 35-percent increase in water productivity (more "crop per drop") could reduce the agricultural share of water consumption from 80 percent to 20 percent in many areas. Supplemental irrigation, groundwater harvesting, and recycled wastewater can be used. Drought resistance can be added to plants via traditional breeding and also by biotechnology – although declining biodiversity, also related to water stress, reduces our chances of finding the right characteristics.

We cannot create water but we can manage it better, much better. Absurdities need to be fixed: many major cities lose over 40 percent of their supply in leaking pipes. In some cities, few bills are sent out and even fewer paid, meaning no funds for maintenance or service extension to low-income areas. Some irrigation systems waste 70 percent of their water.

There is enough water for all our purposes if it is managed properly – and equitably. Poor water management threatens the woman walking to the distant well, the farmer at the end of the irrigation canal, and the slum dweller paying several times as much per litre as a richer citizen. It also threatens the survival of freshwater fish, forests, crops in fields parched by recurring drought, and inland lakes diminishing year by year. New technology helps, but cannot solve these problems. Water scarcity is as much about the inadequacy of financing and management, as of water resources.

Some remedies may be obvious, but few are simple. There are taboos against the re-use of wastewater. Cleaning out corrupt or inept water administrations raises huge political problems. And there are controversies – over genetically modified crops, regulations and their enforcement, the involvement of the corporate sector in water management, the construction of large dams and many other issues. Some "solutions" generate more enemies for governments than friends.

I welcome this wide-ranging Atlas, created by knowledgeable and caring authors, Maggie Black and Jannet King. I know it will inform – but I also hope it will inspire readers to *do something*. We all can. Start by turning off the tap, and talking about it. Only drink bottled water when absolutely necessary. Might a couple of veggie-only days per week improve your health and that of the planet? Get to know your own river. Find out about pollution in your area – who is polluting (it may be you), and what it might take to reduce this. What part of the natural environment is at risk from local water depletion, and how wasteful is your city's water system? Be vocal, and on the right side of difficult areas of national and municipal decision-making.

There are no silver bullets and difficult times are on the horizon. But Water *is* Life; and life itself is at stake.

Margaret Catley-Carlson
Patron of the Global Water Partnership www.gwpforum.org
Chair of the World Economic Forum Agenda Council on Water,
and member of the Secretary General's Advisory Board
on Water and Sanitation.

Introduction

Water means life – a truism so often repeated that its significance becomes lost. This vital natural resource – falling from the sky, bubbling up into springs and lakes, flowing in streams and rivers – is so fundamental to human activity that it has to be managed in such a way that everyone has access. Leaving aside questions of unequal power over resources, the very nature of water militates against this. Rain falls equally "on the just and on the unjust", but everything depends on where they are standing. There is nothing just about annual rainfall distribution, which varies from a few millimetres in some places to thousands in others. And as the climate comes under increasing pressure, the meteorological patterns that scientists have worked for generations to understand are becoming less predictable. Rainfall, which regenerates all other surface and underground sources, may be about to become even more unjust than before.

Rain's erratic choice of landing place, from deserts to forests, tropical to temperate zones, mountains to valleys, is not the end of this complicated story. Unlike other elements on which life depends, it frequently changes its state – from liquid to vapour, from liquid to ice, and *vice versa* depending on the season. Water in lakes and reservoirs is constantly evaporating. And water in the landscape never stands still – it is always on the move. It seeps into the soil for use by plants and creatures, or percolates into aquifers where it renews underground supplies. It travels downhill on even the slightest of gradients. Navigating around whatever impediments it finds in its way, surface water enters a complex system of streams and tributaries connecting river basin to river basin, joining an ever larger flow destined for the sea. Many of these linked river networks are occupied by different peoples, states and jurisdictions.

At any and every stage along its journey, water is used – and sometimes re-used several times – to support life and economic activity. Maximizing its potential for different uses and environments requires technology, investment, control of pollution, regulation, and efficient service delivery. For example, some of the rain falling on the Tibetan Plateau finds its way into China's Lancang river. At many stages downstream, water is diverted by hydraulic construction and human ingenuity into paddy fields to support rice growing. Elsewhere, the flow generates electricity as it is channelled through hydropower turbines. Towns and cities remove water for human consumption and industry, and discharge wastewater back into it. As the river slows and broadens in its lower reaches, becoming the Mekong, it supports a vast aquatic environment on which local fisherman and wildlife depend. Eventually it enters the South China Sea, having passed through six countries and been endlessly manipulated and exploited along the way.

Heightened demand

In an ever more crowded world, the processes involved in this manipulation and exploitation are becoming increasingly complex. Much more is being demanded of hard-pressed water resources. Rivers have been increasingly fragmented by dams. Upstream users are reducing both the volume and the quality of water descending downstream. Non-renewable supplies, in the form of fossil water aquifers formed millennia ago, are becoming rapidly exhausted. Every drop of available supply has to be harnessed to agricultural, industrial or domestic use – and sometimes all three in sequence. The volume of renewable supplies remains constant and is unlikely to falter, even though fluctuations caused by global warming are affecting the performance of the hydrological cycle. But the pressures exerted on this finite supply, both from increased population, and from the increasing numbers of people expecting to enjoy an industrialized lifestyle, are profound. Competition between different types of use, and between upstream and downstream users, is becoming more acute. The extraordinary nature of the substance compounds the many difficulties of managing water in such a way that all these conflicting interests are adjudicated fairly.

Awareness of the critical limits on freshwater supplies has been growing over the past 20 years, alongside more general appreciation of environmental constraints. Indeed, the circulation of water in the environment – to preserve wetlands, conserve biodiversity, and protect climate stability – has itself become recognized as a category of "water use" necessary to nurture the planet and its other life-giving resources. One strategy for water conservation has been to attach an economic value to all its uses and apply market instruments, such as water-pricing, to prevent profligate extraction and consumption. But the treatment of water as a commodity like any other, to be traded and used for corporate profit-making, has caused huge resentment. In those societies where poverty is acute, and rural farmers and urban dwellers are surviving at levels close to subsistence, unsubsidized water services effectively mean no water services at all. However important it is to conserve supplies, the story of water will be even more unjust if the least well-off bear a disproportionate burden of the costs.

In fact, the problem of water as it relates to people in non-industrial environments is that most of them use too little water, rather than too much. Around 1 billion people are still without a reliable source of drinking water, and 2.5 billion people are without sanitation. Having no tap at home constrains water use to the point where lack of personal hygiene is at least as much of a disease risk as lack of safe drinking water. Any attempt to improve water management in such a way as to make distribution more just should spread services to those without, many of whom currently spend much more on water, purchased by the litre and carried home in a pot, than those living where pipes and taps are prolific.

With water, as with pressure on other natural resources, it is not the poor who are pumping up industrial-scale quantities to market it in bottles, or to irrigate sugar or cotton plantations in unsuitable dryland environments. Nor are they manufacturing or buying televisions, computers, cars or other sophisticated consumer products. It is not the disadvantaged and underfed who are polluting rivers with pesticide residues and chemical wastes, or eating farmed fish or hamburgers requiring large quantities of water for their production. The industrial lifestyle is propped up by water, as much as or more than it is propped up by oil.

Water profligacy

Food is one of the most thirsty water consumers. Over 70 percent of water withdrawals are used for agriculture, to flood fields or spray crops. But much of this water fails to reach its target – the roots of the plants; it is lost to the atmosphere, or returned to the water system unused. If poorly managed, irrigation can actually damage the soil, leaving it saline and unproductive. For this and many other reasons, including the social disruption and environmental damage caused by large dams, it is generally acknowledged that the train of "progress", in which large-scale irrigation projects opened up new agricultural land for cultivation, has run into the buffers.

Despite the uneven distribution of land ownership and cheap food, since the expansion of food production that accompanied the Green Revolution it has been possible to envisage a time when no child would go to bed hungry. But over 850 million people are still without a sufficient or nutritious diet, and the day when these figures will improve may now be receding into the future. If food production is to keep pace with increasing population, and prices are to be kept in check, water efficiency in agriculture will have to be given far more attention. Volumes available for agriculture are likely to decline or remain static as industry and expanding urban centres increase their share.

Technology will have to be harnessed to reduce water wastefulness. Up to now, hybrid seeds have mostly required extra water for cultivation; more attention will have to be focused on plant strains that require less water. Farmers will have to rediscover respect for environmental parameters, with drylands used for drought-resistant grains and tubers. Investment is needed in small-scale irrigation and water-harvesting techniques, which could improve the livelihoods of millions of farming families, especially in Africa but also in South Asia. Irrigation needs to be carefully managed, and combined with measures to improve the water-holding quality of the soil, nurture its fertility and increase yields from rain-fed crops. As pressures mount, food production will have to focus on items that use less water per unit of energy or nutrition than the red meat so highly prized in Western cultures.

Reduction of water profligacy and improved efficiency are also needed for water use in manufacturing. In many Western countries, water conservation has been enforced by regulation and pricing to the point that recent expansion in industrial water use has been relatively constrained. The challenge is to ensure that these kinds of measures are taken up by newly industrializing countries where water governance and regulatory frameworks are less developed and more frequently flouted. On the domestic front, appliances such as toilets and washing-machines that use less water are now widely available, but even these, as they are taken up by the new middle-class in countries such as China and India, will have a major impact on the quantity of fresh water used in towns and cities, and on the quantity of wastewater discarded.

The amount of water used per household varies enormously around the world, and a large part of it is invisible. Consumption is not limited to drinking, bathing, flushing the toilet, using the dishwasher and watering the garden, adding up to well over 100 litres per capita a day. Everything that is manufactured – from electronic equipment to newspapers and kitchen gadgets

– has involved water in its production. The total amount of water each person consumes if such products are factored into our "water footprint" is far higher than the figure for direct consumption. Nor is usage restricted to water from local sources: it also includes water embedded in food and goods imported from elsewhere. Thus water-stressed areas in Africa, America, Asia and Australia may be used to produce consumer items for export, while – with real injustice – local farmers and herders go short.

Pollution

At the same time as demand on volumes increase, pressures mount on freshwater quality. No longer can natural water from springs, dug wells and running rivers automatically be assumed to be clean and safe to drink. The natural capacity of waterways to act as the world's inbuilt washing-up apparatus is inadequate to cope with the overload of wastes from increased population density. Many towns and cities in the developing world suffer the indignity of London 150 years ago when, in a hot summer season, the Thames was reduced to a Great Stink by a combination of upstream take-off and raw sewage inflows. Around 90 percent of human waste in the developing world is still discharged untreated into rivers. Since the threat of a cholera epidemic by the intake of foul breath no longer causes the alarm it once did, the public-health incentive for dealing with this nuisance is not what it used to be.

Human waste is an organic substance, and although overload in rivers, lakes and streams destroys plant and aquatic life and presents serious disease problems, the pollution caused by chemical wastes and industrial spills is far more damaging. Failure to regulate or to enforce regulation on polluters who discharge wastes at industrial sites can have disastrous effects on the aquatic environment, killing fish stocks and the ecosystem on which they survive. Some pollutants are not absorbed in water, and may be traced thousands of miles away from their original discharge. Where chemical fertilizer and pesticide residues, or pharmaceutical components, are washed into rivers or leached into the soil, and from there enter the food chain, their toxic effects may build up in human tissues and cause long-term ill-health. For too long, the world's freshwater and seawater network has been considered as having an unlimited capacity to function as humanity's sink. As a result, many parts of the network have become degraded.

Co-operation over water

The increasing pressure on water resources has led to intense competition. Within one community, it is often hard to agree who has the right to take freely from the source for irrigation purposes, or whether people with a tap in their yard should pay a water-rate higher or lower than those still obliged to walk to the pump. Should fines be imposed on people whose tannery, or cloth-dying business, or latrine has fouled the local source? These are questions that have exercised communities for centuries. At the local level, water governance has always demanded co-operation, often reinforced by water's venerated place in human affairs. But as lifestyles become more water-intensive, and the supply is tampered with at ever greater distances, these problems become more acute, especially at the wider level of district or nation, up to multinational level.

Many fear that water is becoming a commercialized commodity, with market forces left to decide who gets to use it or abuse it. Fortunately, that prospect is retreating. Irresponsible profit-making and corruption over water services – the result of inflated expectations from the privatization of services and the efficacy of markets – and the difficulties entailed in persuading customers and authorities to accept much higher pricing regimes, has induced a major re-think about the best distribution of water management between public and private sectors. Compared to the late 1990s, there is now a much wider appreciation that water is a common good, and that it ought to be managed in the common interest, by authorities that are answerable in the public domain. When the task of reconciling all the different user interests is understood in all its parameters, it seems likely that a new business of "water diplomacy" among all sorts of public and private practitioners is going to be a growth industry for the 21st century.

The idea of "integrated water resources management" sounds so reasonable and just – reconciling upstream and downstream users, allocating so much to agriculture and so much to industry, bringing in all parties across all political boundaries to the river basin forum – that it ought to be adopted universally and without delay. But its realization requires a complex process of reconciling competing claims, and a willingness to share a natural resource in an equitable way; such an achievement would be virtually unprecedented in human history. The omens, however, are more positive than might be thought. Despite all the talk about "water wars", experience shows that co-operation over water has occurred more often than conflict, and that antagonists with deeply held differences in almost every sphere can manage to find common cause over water. In the end, the unjust distribution of water in the landscape may provide the stimulus for humanity to find a way of sharing this life-giving resource, and thereby further the cause of peace.

Maggie Black
Oxford

Jannet King
Brighton
April 2009

Acknowledgements

The authors and publishers gratefully acknowledge the help generously given in the form of maps and data by the following individuals and institutions:

The **WHYMAP** team, Bundesanstalt für Geowissenschaften und Rohstoffe (BGR), for use of the groundwater resources map on pages 26–27; **Chris Milly** of the US Geological Survey for providing a map showing projected change in run-off on page 30–31; **Christer Nilsson**, Landscape Ecology Group, Umeå University for the map of river fragmentation on pages 36–37; **Annette Prüess-Üstün** and **James Bartram** of the World Health Organization for making data on diseases available for use on pages 52–53; **Aaron Wolf** and **Lucia DeStefano** of the Program in Water Conflict Management and Transformation, Oregon State University for data used on pages 88–91; **Arjen Y. Hoekstra**, Twente Water Centre www.waterfootprint.org for the data behind the map on pages 94–95.

We would also like to thank the following photographers and organizations who have supplied images:

Page 18 Mike Manzano / iStockphoto; 28 Michael Fuller / iStockphoto; 30 iofoto / iStockphoto; 36 Research worker: International Rivers; Huemul deer: Christian Saucedo, International Rivers; Site of proposed dam in Patagonia: Gary Hughes, International Rivers; 38 Andrea Krause / iStockphoto; 44 UNICEF Zambia; 49 UN-HABITAT; 51 UNICEF Zambia; 54 World Bank / Eric Miller; 58 Susanne Wong / International Rivers; 60 Steve Mcsweeny / iStockphoto; 62 Extreme affordability, Hasso Plattner Institute of Design at Stanford / Klaus http://extreme.stanford.edu ; 71 Canal boat: Graham Heywood / iStockphoto; Artificial beach, Paris: www.aquamedia ; Kerala: Robert Churchill / iStockphoto; Golf course: Sheldon Kralstein; Amazon cruise: www.travelwizard 74 Online Creative Media / iStockphoto; 79 Baltic Sea: http://earthobservatory.nasa.gov ; Gulf of Mexico: http://serc.carleton. edu ; 81 Dead fish: Alan Septoff / Tibor Kocsis media.earthworksaction.org ; 83 Flamingos: Charles Schug / iStockphoto; Otters: John Stezler / iStockphoto; Frog: Samuli Siltanen / iStockphoto; Crocodile: Keiichi Hiki / iStockphoto; Trout: Laurin Johnson / iStockphoto; Water lilies: Isabelle Lewis; 88 World Bank flickr library; 84 World Bank / Tomas Sennett; 92 River scene: Curt Carnemark / World Bank; Building water pump: Curt Carnemark / World Bank; Farmer: Ray Witlin / World Bank; 93 Marco Betti / WaterAid; 97 Cochabamba: www. worldforum.org ; 98 Check dam: Shree Padre; Seawater greenhouse: courtesy Charlie Paton, Seawater Greenhouse Ltd; 100 Kris Hauke / iStockphoto.

We have made every effort to obtain permission for the use of copyright material. If there are any omissions, we apologize and shall be pleased to make appropriate acknowledgement in any future edition.

PART 1 A Finite Resource

Water is fundamental to all life and human activity, and it is under serious threat. This is not because the supply is dwindling significantly. Although some aquifers in dry regions are being exhausted, the renewable supply of water on which planetary survival and well-being depend remains constant. The hydrological cycle lifts this water from seas and releases it as rain or snow to nourish plant growth and fill rivers, lakes and underground aquifers. And even though the behaviour of some parts of the cycle may alter due to climate change, this is not the most immediate threat to our global water resource.

The problem lies in excessive demand. An increasing number of people, and especially of those able to afford water-profligate lifestyles, are stretching the fixed available supply to its limits. People with industrialized standards of living consume diets rich in foodstuffs requiring extra water to produce, and demand goods such as cars, television sets and computers that require water for their manufacture. On top of this comes pressure on water quality. The waterways that constitute our natural drainage system are unable to absorb increasing loads of human waste, nor the industrial and agricultural chemical effluents that end up polluting rivers, lakes, and coastal zones.

The ability to manipulate water and manage its use for productive purposes has always been central to human development. Leaders of the ancient world depended on hydraulic works – dams, lifting devices and artificial lakes – to develop and maintain their civilizations. They fully understood the variability of rainfall and run-off within their domains, and its implications for populations in rain-short areas, long before today's pressures had to be taken into account. Renewable supplies are not distributed evenly around the world or within its different topographical regions – even between the hills and valleys of a contiguous zone.

Those in water-short areas have up to now tackled their problems by capturing run-off behind dams, and storing or diverting water for agricultural or other uses. As more and more water is manipulated in this way, the environmental and other limits of this approach have become apparent. Upstream and downstream parties to the same resource, in river basin, watershed or groundwater network, are forced into an ever fiercer competition as populations increase and lifestyles and modes of production alter.

Whatever their uneven distribution, the contents of the global water pot are going to have to suffice for all our needs. New ways of managing water will have to be found in order to maintain quantity and quality and to achieve a fair distribution of this essential resource.

1 THE GLOBAL WATER POT

The volume of water in the world never changes, but only 2.5 percent is fresh, and more than two-thirds of this is unavailable for human use.

There are approximately 1,386 million cubic kilometres of water on the planet. Nearly all of it is salt water, contained in the oceans, seas, saltwater lakes and in aquifers beneath the oceans. Of the 2.5 percent that is fresh water, more than two-thirds is locked up in glaciers, snow, ice and permafrost. Of the fresh water that is technically "available" for people to use, only a tiny proportion is on the surface of the Earth. The rest is to be found underground, in aquifers.

The Earth's water is in constant motion. It is evaporated from the land and oceans by the sun's heat, which turns liquid water into water vapour. Moisture held in vegetation is also lost to the air through the process of evapo-transpiration. In the atmosphere, water vapour condenses into the droplets from which clouds are formed and eventually falls as rain. The key to our survival is that although fresh water flows from the land into the oceans, some of the evaporation from the oceans falls on land, feeding the rivers, watering the soil and restocking the underground aquifers.

Climate change is predicted to have an impact on our freshwater resources and ecosystems, including melting ice-sheets and alterations in rainfall patterns, but this is likely to take place at a local level. The total amount of water available to the planet is not predicted to change.

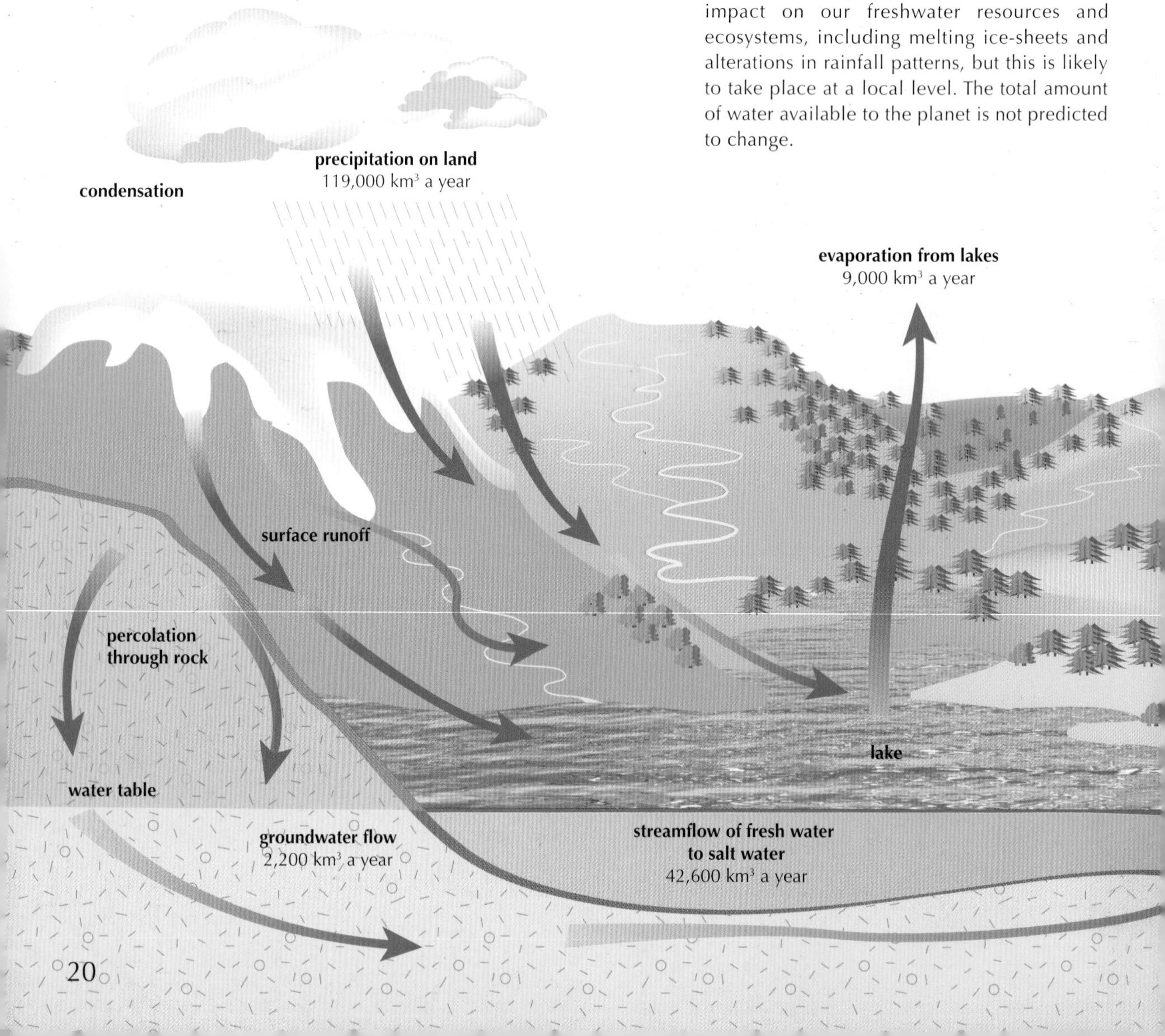

precipitation on land
119,000 km³ a year

condensation

evaporation from lakes
9,000 km³ a year

surface runoff

percolation through rock

water table

lake

groundwater flow
2,200 km³ a year

streamflow of fresh water to salt water
42,600 km³ a year

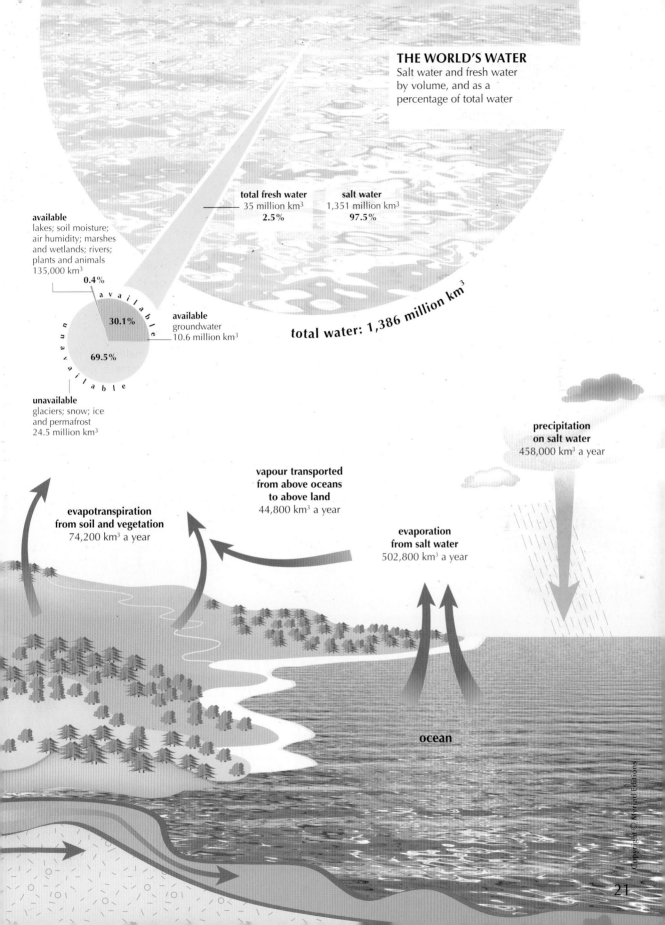

THE WORLD'S WATER

Salt water and fresh water by volume, and as a percentage of total water

total fresh water
35 million km³
2.5%

salt water
1,351 million km³
97.5%

available
lakes; soil moisture;
air humidity; marshes
and wetlands; rivers;
plants and animals
135,000 km³

0.4%

a v a i l a b l e

30.1%

available
groundwater
10.6 million km³

u n a v a i l a b l e

69.5%

unavailable
glaciers; snow; ice
and permafrost
24.5 million km³

total water: 1,386 million km³

**precipitation
on salt water**
458,000 km³ a year

**vapour transported
from above oceans
to above land**
44,800 km³ a year

**evapotranspiration
from soil and vegetation**
74,200 km³ a year

**evaporation
from salt water**
502,800 km³ a year

ocean

2 WATER SHORTAGE

California, USA

Following a third year of drought, the governor of California declared a state of emergency in Feburary 2009, giving him powers to implement water rationing. He called on the water authorities to reduce the economic impact of the drought by conserving and transferring water and by encouraging efficient irrigation practices, and on urban dwellers to reduce their water use by a fifth.

SHORTAGES EXPECTED
by state water managers
in the USA *2008*

- statewide
- regional
- local
- uncertain
- none
- no response

By 2025 nearly

2

billion

people will be living in water-short regions

Global water shortage is not about a reduction in the total supply. The amount of fresh water that lands as rain, filling streams, rivers and lakes, remains constant at 12,500 cubic kilometres, and at present we are still using less than a third.

The problem lies in the mismatch between where the rain falls, and where people live. In many densely populated places, the renewable water resource is insufficient, leading to water being extracted from rivers and underground aquifers at an unsustainable rate. Increasing populations, expanding cities, and the swelling number of those enjoying a water-rich lifestyle are combining to cause critical localized shortages.

Another major cause of water shortage is its heavy use in irrigated agriculture in areas that are dry, or where rainfall is concentrated in seasonal downpours. Some regions, including the Indo-Gangetic plain in South Asia, North China, and the High Plains in North America, are dependent on storing and re-distributing water for agriculture.

Large parts of the African continent are prone to drought and one-third of the population – 300 million – live in conditions of water scarcity. In many areas, the potential for storing and distributing surface water by use of dams and infrastructure is inhibited by hostile geographical and other limiting factors.

Brazil

The Amazon region receives nearly 75% of Brazil's water, but is very lightly populated. The northeast coastal region, where 20% of people live, receives only 2%.

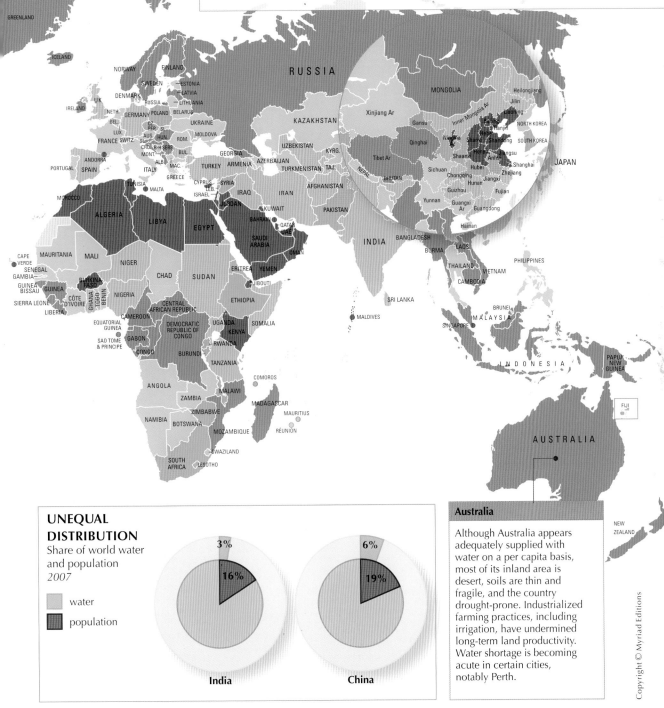

WATER SHORTAGE

Annual renewable water resources
latest available data 2003–07
cubic metres per capita

- fewer than 1,000 *water scarcity*
- 1,000 – 1,699 *water stress*
- 1,700 – 2,999 *insufficient water*
- 3,000 – 9,999 *relatively sufficient*
- 10,000 or more *plentiful supplies*
- no data

UNEQUAL DISTRIBUTION

Share of world water and population
2007

- water
- population

India
3%
16%

China
6%
19%

Australia

Although Australia appears adequately supplied with water on a per capita basis, most of its inland area is desert, soils are thin and fragile, and the country drought-prone. Industrialized farming practices, including irrigation, have undermined long-term land productivity. Water shortage is becoming acute in certain cities, notably Perth.

3 RISING DEMAND

1900:

350

cubic metres

2000:

642

cubic metres
**of water
used per capita
each year**

Around 4,000 cubic kilometres of fresh water are withdrawn every year – equivalent to roughly 1,700 litres per person per day.

Although this is more than anyone needs for personal use, even to fill their swimming pool and sprinkle their garden around the clock, we consume a large amount of water indirectly, embedded in food and industrial products. Meat-rich diets and other attributes of high-consumption lifestyles, such as the acquisition of cars, television sets, and goods whose manufacturing processes require water, absorb ever larger quantities.

Thus, the rapid rise in demand, experienced across all categories of water use (agricultural, industrial, and domestic/municipal), is a reflection of changing lifestyles and of rising numbers enjoying them. Domestic use – for drinking, bathing, cleaning, and for water in offices and non-industrial urban spaces such as parks – appears modest compared to that used for agriculture and industry. But it is consumption of embedded water that fuels rising demand across the board – promoting economic growth, and representing the "development" to which all aspire. Those denied development make the lowest demands, often using fewer than 25 litres per person for all purposes.

Water for agriculture is by far the largest extractive category. This is because of lower industrialization in developing regions, and the dependence of certain tropical areas on irrigation from rivers, reservoirs and aquifers. Most irrigation systems are hugely water-profligate. In places where agriculture is mainly rain-fed (temperate zones and much of Africa), water for domestic consumption may also be used to water plots, gardens and livestock. In a few countries where there is very little industry or irrigation, this makes it appear that more water is used for domestic purposes than for agriculture, even though people's entire subsistence is derived from the land and natural resource base.

WORLD WATER USE
by sector
2001

10%
domestic

20%
industry

70%
agriculture

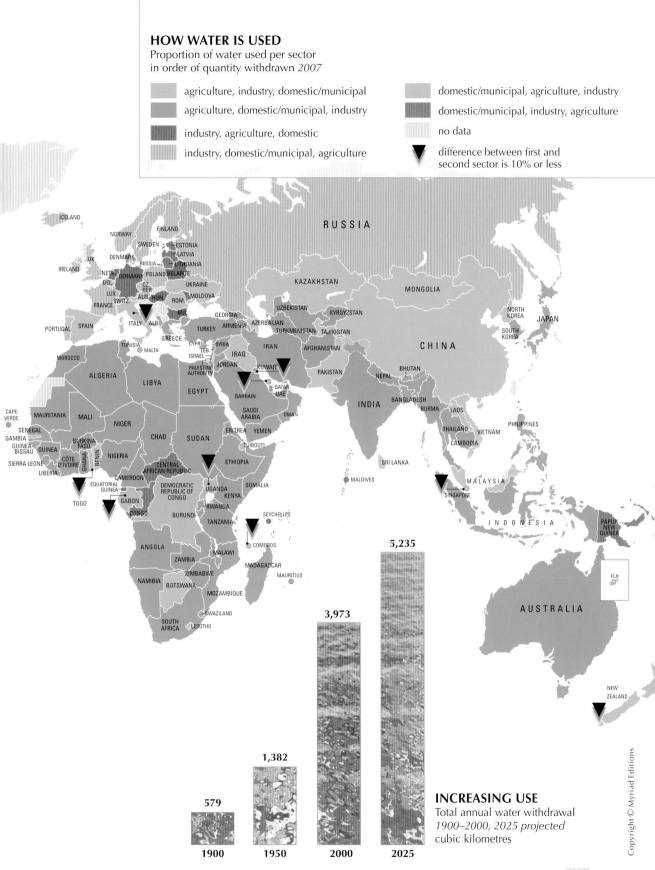

HOW WATER IS USED

Proportion of water used per sector
in order of quantity withdrawn *2007*

- agriculture, industry, domestic/municipal
- agriculture, domestic/municipal, industry
- industry, agriculture, domestic
- industry, domestic/municipal, agriculture
- domestic/municipal, agriculture, industry
- domestic/municipal, industry, agriculture
- no data
- ▼ difference between first and second sector is 10% or less

INCREASING USE

Total annual water withdrawal
1900–2000, 2025 projected
cubic kilometres

1900	1950	2000	2025
579	1,382	3,973	5,235

4 DWINDLING SUPPLY

1.5 billion
people rely on groundwater for their survival

800 cubic kilometres
of water is extracted from the ground each year

About a fifth of water used comes from aquifers. Some of these underground stores are replenished as rainwater seeps through soil and rock, but others, in areas of low rainfall, are non-renewable, and many of these "fossil-waters" are being irreversibly mined.

The volume of water stored in aquifers is hard to assess, and data on its use is patchy, but in many parts of the world it is being withdrawn recklessly. This is because of its increasing use in recent years for irrigation – a voracious consumer of water. Comparatively cheap pumping technology and negligible charges and regulations have led to over 20 million tube wells being sunk by farmers in India alone. The country now withdraws five times as much groundwater as in the 1960s, leading to a significant drop in water tables. This makes it harder to reach new sources, while existing wells constantly run dry and have to be deepened or replaced.

Similarly, in North Africa and Arabia, aquifers created during the last ice age, when the region was very different from the arid terrain of today, are rapidly being sucked dry. Libya's "Great Man-made River" has cost $32 billion and will last only 15 to 20 years.

Many cities rely heavily on groundwater, which is preferred to surface water for drinking because it is less likely to be contaminated. However, exhaustive withdrawals can lead to serious land subsidence, and in coastal areas salt water may intrude, turning the water brackish and unusable.

Groundwater can no longer be regarded as an unlimited supplement to surface water sources, and its sustainable management, including aquifer recharge, is now an important focus of initiatives in the developing world. In some water-scarce regions, particularly in rural areas, there has been a profusion of small-scale efforts at groundwater recharge by use of rainwater harvesting and other conservation methods.

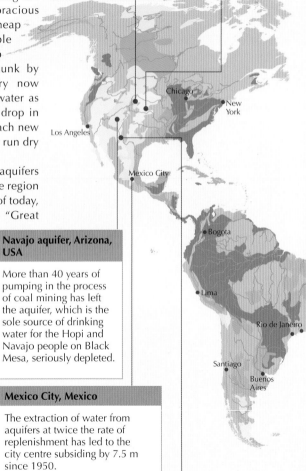

Denver, Colorado, USA
The number of wells in the Denver Basin aquifer system nearly trebled between 1985 and 2001, and the level of the aquifer declined by 76 m. There is concern that the groundwater supplies may be essentially depleted in 10 to 15 years in areas on the west side of the basin.

Great Plains, USA
The level of the Ogallala aquifer, which spans eight states and supplies around 25% of the US irrigated farmland, has fallen by an average of 4 m since the 1950s, and by 30 m in western Texas.

Navajo aquifer, Arizona, USA
More than 40 years of pumping in the process of coal mining has left the aquifer, which is the sole source of drinking water for the Hopi and Navajo people on Black Mesa, seriously depleted.

Mexico City, Mexico
The extraction of water from aquifers at twice the rate of replenishment has led to the city centre subsiding by 7.5 m since 1950.

Mexico–USA border
The aquifer that supports the two cities of Ciudad Juarez and El Paso is becoming increasingly brackish, and is predicted to run dry by 2025 unless it can be successfully managed.

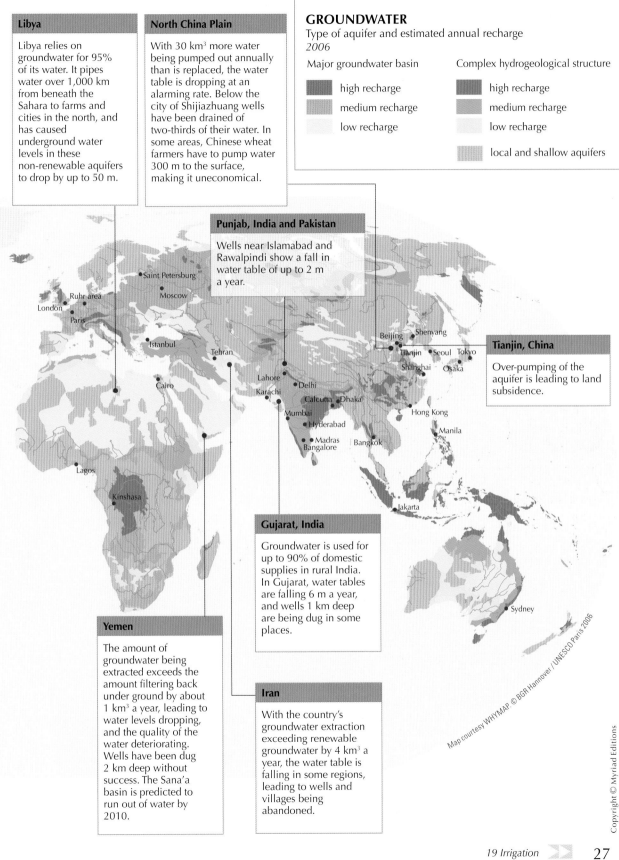

Libya

Libya relies on groundwater for 95% of its water. It pipes water over 1,000 km from beneath the Sahara to farms and cities in the north, and has caused underground water levels in these non-renewable aquifers to drop by up to 50 m.

North China Plain

With 30 km³ more water being pumped out annually than is replaced, the water table is dropping at an alarming rate. Below the city of Shijiazhuang wells have been drained of two-thirds of their water. In some areas, Chinese wheat farmers have to pump water 300 m to the surface, making it uneconomical.

GROUNDWATER

Type of aquifer and estimated annual recharge
2006

Major groundwater basin

- high recharge
- medium recharge
- low recharge

Complex hydrogeological structure

- high recharge
- medium recharge
- low recharge
- local and shallow aquifers

Punjab, India and Pakistan

Wells near Islamabad and Rawalpindi show a fall in water table of up to 2 m a year.

Tianjin, China

Over-pumping of the aquifer is leading to land subsidence.

Gujarat, India

Groundwater is used for up to 90% of domestic supplies in rural India. In Gujarat, water tables are falling 6 m a year, and wells 1 km deep are being dug in some places.

Yemen

The amount of groundwater being extracted exceeds the amount filtering back under ground by about 1 km³ a year, leading to water levels dropping, and the quality of the water deteriorating. Wells have been dug 2 km deep without success. The Sana'a basin is predicted to run out of water by 2010.

Iran

With the country's groundwater extraction exceeding renewable groundwater by 4 km³ a year, the water table is falling in some regions, leading to wells and villages being abandoned.

Map courtesy WHYMAP. © BGR Hannover / UNESCO Paris 2006

Copyright © Myriad Editions

5 COMPETITION AND CONFLICT

Around

260

river basins

**are shared by
two or more
countries**

As populations grow, and more water is extracted per person, there is increasing competition and conflict over the exploitation of river waters and aquifers.

Many countries depend for much of their water supply on rivers that flow in from another country. When rivers are dammed or their flows diverted, this leads to potential conflict. The discharge of pollutants can also pit downstream against upstream inhabitants.

These pressures have led to talk of "water wars". However, although the manipulation of water has played a part in hostilities – including the draining of Iraq's southern marshlands in 1993 by Saddam Hussein, which removed the means of survival from 500,000 people – wars have yet to be explicitly fought between nations over water. There have been military and terrorist actions to destroy dams, cut off supplies or pollute water sources, but these form part of wider campaigns to disrupt economic life or pursue a political cause. Other conflicts over water derive from territorial disputes.

Water supplies do feature strongly in some major political disputes, including the Israel-Palestine conflict. In Central Asia, the collapse of the USSR led to confrontation between six successor republics over a previously centralized dam and irrigation network. In India, a dry year can lead to inter-state violence over inadequate water releases by upstream states, with sit-ins, riots and attacks on vehicles crossing state borders.

Competition over water within a country, between groups with conflicting interests, is not uncommon. Industrial users and farmers may dispute use of scarce resources, or companies with a commercial interest in water supplies may find themselves at loggerheads with local people who rely on the same water for their survival.

Actual water-related confrontations, such as dam-site captures and sabotage are typically localized. These disputes are increasing, but have to be solved at national or international level – through the many tribunals and river basin organizations that now exist.

The Nile basin

The headwaters of the Nile are vital for existing livelihoods and future development in upstream countries such as Ethiopia and Kenya, but under colonial-era agreements, use of the Nile is controlled by Egypt and Sudan. The internationally sponsored Nile Basin Initiative, involving all 10 basin countries, is trying to establish an acceptable framework for Nile waters management and distribution, while political tensions increase.

Bolivia–Chile

The Silala river flows from Bolivia into Chile, where it is vital to the processing of copper in the Atacama desert. A disagreement over the right to its waters is part of a long-running territorial dispute between the two countries. Bolivia claims that the river's current channel is "artificial", and that Chile therefore has no claim to the water. In 2006, President Evo Morales announced plans to "industrialize" the water, reducing the flow to the mines.

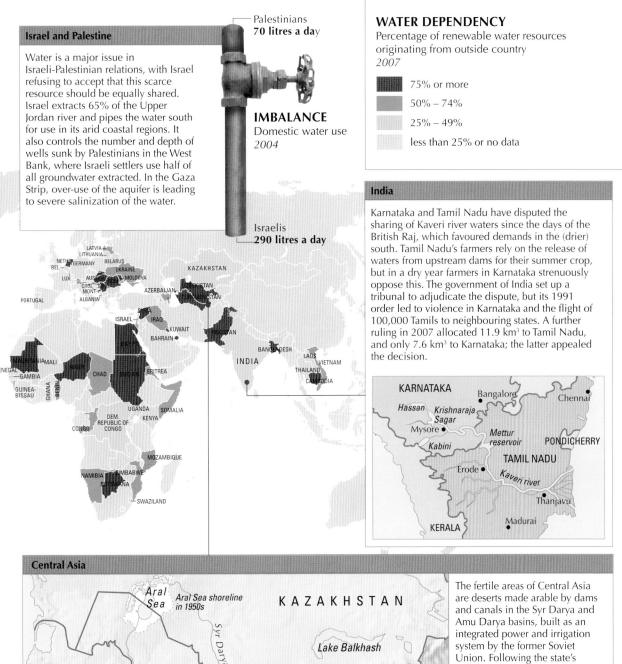

Israel and Palestine

Water is a major issue in Israeli-Palestinian relations, with Israel refusing to accept that this scarce resource should be equally shared. Israel extracts 65% of the Upper Jordan river and pipes the water south for use in its arid coastal regions. It also controls the number and depth of wells sunk by Palestinians in the West Bank, where Israeli settlers use half of all groundwater extracted. In the Gaza Strip, over-use of the aquifer is leading to severe salinization of the water.

Palestinians
70 litres a day

IMBALANCE
Domestic water use
2004

Israelis
290 litres a day

WATER DEPENDENCY
Percentage of renewable water resources originating from outside country
2007

- 75% or more
- 50% – 74%
- 25% – 49%
- less than 25% or no data

India

Karnataka and Tamil Nadu have disputed the sharing of Kaveri river waters since the days of the British Raj, which favoured demands in the (drier) south. Tamil Nadu's farmers rely on the release of waters from upstream dams for their summer crop, but in a dry year farmers in Karnataka strenuously oppose this. The government of India set up a tribunal to adjudicate the dispute, but its 1991 order led to violence in Karnataka and the flight of 100,000 Tamils to neighbouring states. A further ruling in 2007 allocated 11.9 km³ to Tamil Nadu, and only 7.6 km³ to Karnataka; the latter appealed the decision.

KARNATAKA
Bangalore
Chennai
Hassan
Krishnaraja Sagar
Mysore
Kabini
Mettur reservoir
PONDICHERRY
TAMIL NADU
Erode
Kaveri river
Thanjavu
Madurai
KERALA

Central Asia

Aral Sea
Aral Sea shoreline in 1950s
KAZAKHSTAN
Syr Darya
Lake Balkhash
UZBEKISTAN
reservoir lake Toktogul
TURKMENISTAN
Amu Darya
reservoir lake Shardara
Tashkent
Bishkek
Ashkhabad
Karakum Desert
Bukhara
Ferghana Valley
KYRGYZSTAN
Karakum Canal
IRAN
reservoir lake Nurek
Dushanbe
proposed dams
CHINA
proposed artificial Golden Century Lake
TAJIKISTAN
AFGHANISTAN
PAKISTAN

The fertile areas of Central Asia are deserts made arable by dams and canals in the Syr Darya and Amu Darya basins, built as an integrated power and irrigation system by the former Soviet Union. Following the state's collapse, management of the basins was split between the successor republics. Water- and energy-sharing agreements broke down, and over-use and mismanagement prevailed, severely depleting the Aral Sea. Drought in 2007–08 further reduced upper-reach reservoirs, and led to flare-ups between cross-border communities in the Ferghana Valley.

LATVIA
LITHUANIA
NETH. GERMANY BELARUS
BEL. LUX.
SL. AUS. SLOV. ROM. UKRAINE MOLDOVA
PORTUGAL CRO. MONT. SER.
ALBANIA
AZERBAIJAN UZBEKISTAN
KAZAKHSTAN
ISRAEL SYRIA IRAQ TURKMENISTAN
KUWAIT
BAHRAIN PAKISTAN
EGYPT
MAURITANIA MALI NIGER
SENEGAL GAMBIA CHAD SUDAN ERITREA
GUINEA-BISSAU GHANA BENIN
UGANDA SOMALIA
CONGO DEM. REPUBLIC OF CONGO KENYA
MOZAMBIQUE
NAMIBIA ZIMBABWE
BOTSWANA
SWAZILAND
BANGLADESH
INDIA
LAOS VIETNAM
THAILAND
CAMBODIA

PART 2 Environmental Pressures

Water plays a central role in the protection and health of our environment, from the level of atmospheric and meteorological forces, down to the intimate context of homes and communities. The role water was expected to play used to be closely adapted to the volume available in any given environment, but industrialization is playing havoc with that balance in many parts of the world.

The pressures exerted on our environmental fabric – on cultivable land, living space, freshwater fish and other aquatic life, forests and livestock herds – by expanding populations, changing patterns of human settlement, and consumer lifestyles have inevitable environmental repercussions. The construction of a reservoir to supply an expanding city population, for example, not only has an environmental impact on its site, but upstream and downstream as well, and may even produce greenhouse gas emissions.

Global warming has many water-related impacts. Catastrophic droughts and floods – disasters that used to be seen purely as the product of natural forces – are becoming more common. Climate change is having an impact on rainfall patterns and surface water flows, including the melting of glaciers, rises in sea levels, and increased frequency of extreme weather events of all kinds. But other human factors relating to the management and use of water resources also feed into environmental stress.

The harnessing of rivers behind dams, in order to use their contents for agriculture and other purposes, has fragmented natural water courses to a disturbing extent. Some rivers have now become artificial systems of linked lakes, while so much water is diverted from others that barely a trickle reaches the sea. These alterations of natural flows have led to losses of plant and fish species and floodplain fertility, and other forms of damage to upstream and downstream ecosystems. Until recently, the real costs of such losses – to local environments and people's livelihoods – have been under-appreciated, and are often still ignored.

Similarly, the draining of wetlands used to be regarded as a beneficent process of converting "wet deserts" and "malarial swamps" into land useful for agricultural production and human settlement. The destruction of wildlife habitats, the loss of water for evaporation and hence a reduction in rainfall, and other unforeseen environmental fall-outs have sometimes proved devastating. The need to preserve the natural environment, not only for its beauty but for reasons of long-term human survival and well-being, are now better understood.

Environmental protection and freshwater conservation are two sides of the same coin, and the critical issue facing water policy-makers is how the protection of fresh water and the environment are to be balanced with mounting demands.

6 CLIMATE CHANGE

By 2080 up to

20%

of people will live in areas with increased flood risk

POPULATION IN WATER-STRESSED RIVER BASINS
Based on climate-change predictions
1995 & 2050 projected

5 billion

1.6 billion

1995   **2050**

Climate change is expected to affect rainfall, river flow and freshwater supplies in many and complex ways, and the negative impacts will outweigh any likely benefits.

The predicted increase in global temperatures will influence the whole hydrological cycle, affecting both when, and how much, rain falls. Areas in northern latitudes are likely to receive more rain, and those in mid-latitudes less. The effect of these changes will both exacerbate environmental stresses and increase tensions already created by population growth and urbanization in many overstretched river basins. Across the world, episodes of torrential rain are expected to become more frequent, heightening the risk of floods.

Rising temperatures are also expected to continue a process already begun in the world's mountain ranges – the melting of ice-caps and the retreat of glaciers. Changes in the timing and volume of water flowing from the mountains are likely to have a major impact on the long-term future of the 1 billion people whose survival depends on this seasonal flow. Initially, river flow is likely to increase as a result of the additional melt-water, but if the ice vanishes completely, many of the world's mighty rivers will be severely diminished.

Climate change is also likely to worsen the quality of the world's fresh water. Flooding washes sediment into rivers, bringing with it pathogens, residue from pesticides, heavy metals from industry, and phosphorous, which, combined with warmer water, will result in algal blooms. In urban areas, floods also cause sewage systems to overflow, contaminating the streets and infiltrating drinking-water supplies. Meanwhile, reduced river flows will concentrate pollutants in the remaining water.

All these effects are likely to have negative impacts on the health not only of people, but of entire ecosystems. In addition, as warmer oceans expand, the sea-level will rise, not only inundating land, but introducing salt into freshwater coastal aquifers.

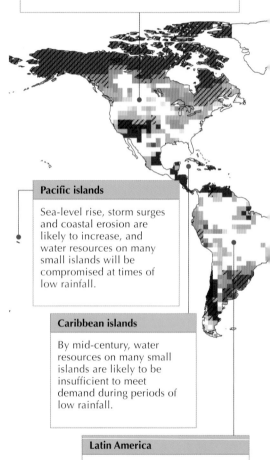

North America
The summer flow of rivers, such as the Columbia, which depend on snowmelt will decline by mid-century. The Colorado river flow is also predicted to decline, threatening the water and electricity supplies of 30 million people and countless businesses. Recharge of the Ogallala aquifer, which supplies a quarter of the USA's irrigated land, is also compromised by the prospect of temperature rise. The effect on water levels in the Great Lakes is uncertain, but an increase in water temperature will impair water quality.

Pacific islands
Sea-level rise, storm surges and coastal erosion are likely to increase, and water resources on many small islands will be compromised at times of low rainfall.

Caribbean islands
By mid-century, water resources on many small islands are likely to be insufficient to meet demand during periods of low rainfall.

Latin America
Tropical forest in the eastern Amazon is likely to be transformed into savannah as temperatures increase and moisture is lost from the soil. Changes in rainfall and the disappearance of glaciers will affect the volumes of water available for consumption, agriculture and power generation. Peru's arid coastal plain is especially vulnerable.

2 Water Shortage; 5 Competition and Conflict

Europe

Reduced snow cover and accelerated glacial retreat are likely in mountain areas, with consequences for flora and fauna and winter tourism. In Southern Europe, already experiencing intermittent water shortages, the situation will worsen, leading to agricultural losses and weaker power generation. Flash floods will become more common, as will the risk of heat waves and wildfires.

CHANGE IN RUN-OFF

Projected large-scale changes in annual run-off based on 12 computer models
2090–99 relative to 1980–99

Decrease
- 40% or more
- 20% – 39%
- 10% – 19%
- 5% – 9%
- 2% – 4%
- little change

Increase
- 2% – 4%
- 5% – 9%
- 10% – 19%
- 20% – 39%
- 40% or more

fewer than 66% of models agree

more than 90% of models agree

The collapse of the Asian "water tower"

The mountains of Asia store water in the form of ice and snow. The spring melt supplies seven of Asia's great rivers, which between them support 2 billion people in vast ecosystems. But the ice is melting. The area of glaciers in the Qinghai–Tibet plateau, the source of China's Huang and Yangtze rivers, has been shrinking by 7% a year and, at this rate, two-thirds of China's glaciers will have gone by 2060. The Gangotri glacier, one of the main sources of the Ganges basin, is retreating by 23 metres a year. Without it, the Ganges will be reduced to one-third of its current summer flow, leaving 500 million people short of food and water.

Central and West Asia

Pressure on water supplies will grow as drylands become even more arid. Already water-stressed populations are likely to find supplies decreasing and competition increasing. In Syria, the amount of renewable water is likely to halve by 2025.

South, South-East and East Asia

Although precipitation will probably increase in some parts of Asia, more may fall as torrential rain, making it difficult to capture and manage. Glacier melt will affect flows in rivers, threatening the water supplies of hundreds of millions of people. The Ganges–Brahmaputra delta, occupied by over 100 million people, will be seriously affected by sea-level rise. Diseases associated with water are likely to increase.

Australia and New Zealand

Parts of Australia and New Zealand are projected to receive less rain. This will have a serious impact on food production in Australia, although in New Zealand increases in other parts are expected to offset losses.

Africa

By 2020, an additional 250 million people in Africa could be experiencing increasing water stress, with yields from rain-fed agriculture reduced by as much as 50%.

7 URBANIZATION

Urban population

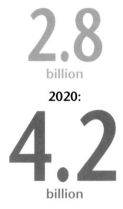

2000:

2.8
billion

2020:

4.2
billion

Sometime in 2008, the world became predominantly urban. With more inhabitants living in towns and cities than in the countryside, huge pressures are exerted on municipal infrastructure and service provision.

Urban growth in developing regions has been extremely rapid, and although some progress is being made worldwide in providing urban dwellers with improved access to water and sanitation services, the pace of growth in most cities means that authorities are constantly playing catch-up. Existing infrastructure is already inefficient, with one-third of the water supply in many systems being lost to leakage, and there may be problems with bringing sufficient water into the city. Many cities have already exhausted nearby surface and underground sources. Amman, Delhi, Santiago and Mexico City are among those pumping water from increasing distances and up increasing heights.

An important feature of contemporary urban growth is that much of it takes place in informal slums and shanty-towns on the fringes of the towns and cities. The proportion of urban dwellers living in such settlements – known variously as *favelas, bidonvilles, bustees*, and simply townships – is typically between 30 percent and 40 percent in

developing regions and can be much higher. Most of these crowded settlements do not enjoy either water or sanitation services, both of which are essential for decent living and health. Squatter citizens are not counted in surveys as their presence in town is "illegal", so the rosy picture typically conveyed by high tallies of urban service coverage is misleading. The implication of the statistics is that the crisis of inadequate water and sanitation, and therefore ill-health, is much more acute in the countryside. This overlooks the reality that in most rural settings there are usable natural sources – wells and streams – and that those living in congested urban environments are more at risk from epidemics. Recent surveys in poor urban neighbourhoods have also found that water and sanitation coverage is significantly worse than reported.

Slum dwellers pay high prices for informal and inadequate water provision, usually from vendors, and most have to manage without toilets. The neglected crisis of urban squalor is growing, while the costs and complications of service provision consistently mount. Despite claims that privatization of utilities would expand service spread to poorer citizens, after a brief honeymoon this has usually failed: the rise in service fees required to make operations fully cost-efficient turns out to be politically unacceptable.

INCREASING URBAN POPULATIONS
1990–2020 projected

👤 10 million urban dwellers

👤 10 million slum dwellers

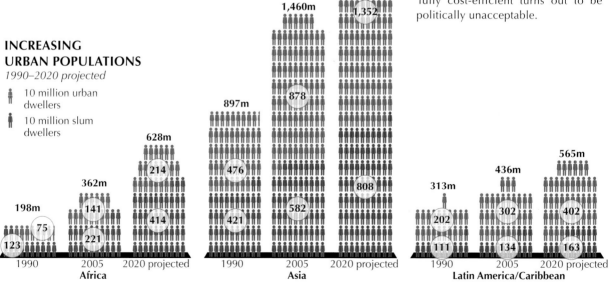

Africa: 1990 198m (123, 75); 2005 362m (221, 141); 2020 projected 628m (414, 214)

Asia: 1990 897m (421, 476); 2005 1,460m (582, 878); 2020 projected 2,160m (808, 1,352)

Latin America/Caribbean: 1990 313m (111, 202); 2005 436m (134, 302); 2020 projected 565m (163, 402)

◀◀ *4 Dwindling Supply*

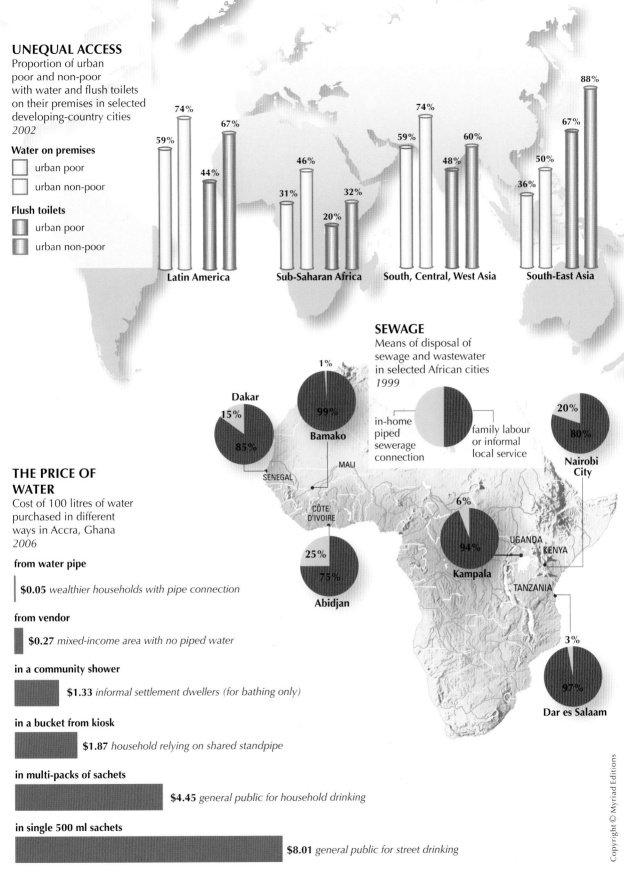

UNEQUAL ACCESS

Proportion of urban
poor and non-poor
with water and flush toilets
on their premises in selected
developing-country cities
2002

Water on premises

☐ urban poor

☐ urban non-poor

Flush toilets

☐ urban poor

☐ urban non-poor

59% 74% 44% 67% — **Latin America**

31% 46% 20% 32% — **Sub-Saharan Africa**

59% 74% 48% 60% — **South, Central, West Asia**

36% 50% 67% 88% — **South-East Asia**

SEWAGE

Means of disposal of
sewage and wastewater
in selected African cities
1999

in-home
piped
sewerage
connection

family labour
or informal
local service

Dakar 15% 85% — SENEGAL

Bamako 1% 99% — MALI

Nairobi City 20% 80%

Abidjan 25% 75% — CÔTE D'IVOIRE

Kampala 6% 94% — UGANDA / KENYA

Dar es Salaam 3% 97% — TANZANIA

THE PRICE OF WATER

Cost of 100 litres of water
purchased in different
ways in Accra, Ghana
2006

from water pipe

$0.05 *wealthier households with pipe connection*

from vendor

$0.27 *mixed-income area with no piped water*

in a community shower

$1.33 *informal settlement dwellers (for bathing only)*

in a bucket from kiosk

$1.87 *household relying on shared standpipe*

in multi-packs of sachets

$4.45 *general public for household drinking*

in single 500 ml sachets

$8.01 *general public for street drinking*

RESERVOIR EMISSIONS

Comparison of range of greenhouse gas emissions by types of power plant
2002
CO_2-equivalent per kilowatt hour

hydro (in tropics) 200–3,000	coal (modern) 1,000	heavy oil 710	natural gas 610

Research on emissions of methane and carbon dioxide from reservoirs is still in its infancy. Field research at the Petit Saut dam, French Guiana (below), suggests that water flowing out of dams may also emit substantial amounts of methane.

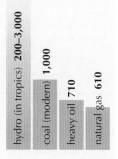

Nearly 60 percent of the world's major rivers are impeded by large dams. During the past century, hydraulic engineering projects have impounded, stored and re-channelled the contents of rivers in a monumental exercise to re-design natural flows.

Since time immemorial, political leaders have harnessed rivers so as to use their contents rather than let them dissipate naturally into the sea. Their purposes include diversion for irrigation, storage in reservoirs, transport and navigation, flood control, land reclamation and, more recently, hydropower. Some works of ancient civilizations were grand and sophisticated, but hydraulic manipulation of rivers reached new dimensions in the past 60 years.

Today, there are 50,000 large dams (over 15 metres high), of which 300 are regarded as giant. This has had the effect of fragmenting many rivers, blocking their natural flow and turning them effectively into series of lakes. On some rivers, the volume of water that can be stored in reservoirs exceeds the annual flow, at times leaving very little water to reach the sea.

The upstream effects of impounding water on such a scale include the destruction of whole ecosystems and the species that depend on them. Human homes, settlements and historic sites, are also submerged. The benefits gained from this destruction may be short-lived. Inundated vegetation rots and releases methane – a greenhouse gas. Sediment, previously carried down the river, falls to the bottom as the water slows, substantially reducing the amount of water that can be held in the reservoir.

Downstream, the loss of the sediment diminishes floodplain fertility. Interrupted or reduced flows destroy wetlands and leave insufficient water for irrigation. Fisheries are also disrupted, with migrating species unable to pass up or downstream. River fragmentation is a major factor in the threat to a third of endangered freshwater fish species.

Mississippi Delta

The damming and embankment of the river over many years has led to the silt it carries being deposited before it reaches the sea. The subsequent erosion of its delta makes New Orleans more vulnerable to flooding.

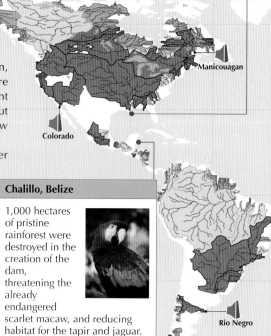

Manicouagan

Colorado

Chalillo, Belize

1,000 hectares of pristine rainforest were destroyed in the creation of the dam, threatening the already endangered scarlet macaw, and reducing habitat for the tapir and jaguar.

Rio Negro

Patagonian rivers

Chile plans to build five hydroelectric dams on two untouched rivers: the Pascua and the Baker. 5,900 hectares will be flooded, including prime agricultural land and one of the world's rarest forests. The electricity generated will be transported to cities and industries, requiring that a 1,600-km corridor be cut through the forest. The environmental impacts are likely to increase the threat to the endangered huemul deer.

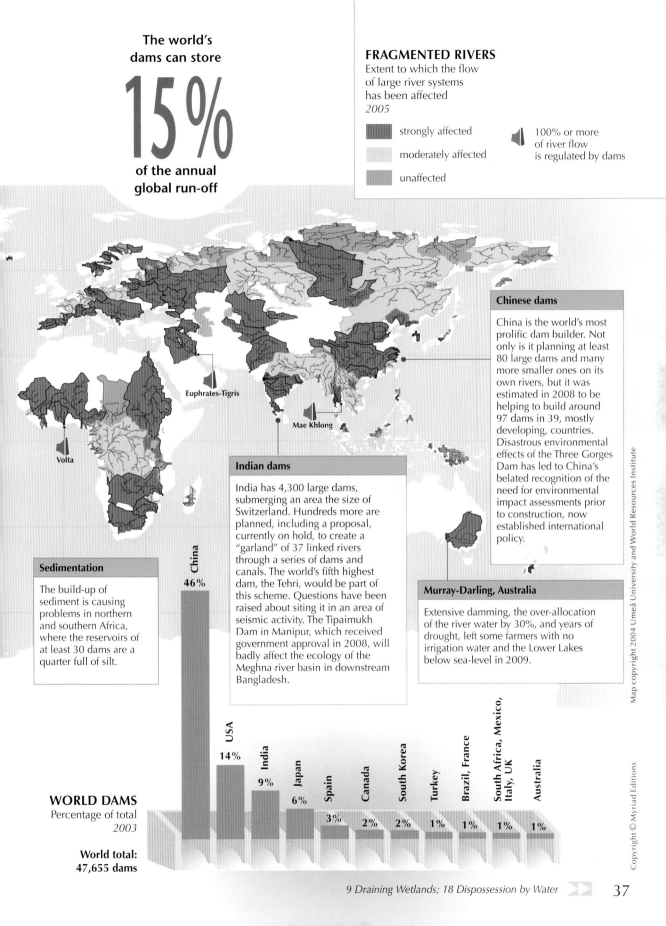

The world's dams can store

15%

of the annual global run-off

FRAGMENTED RIVERS

Extent to which the flow of large river systems has been affected
2005

- strongly affected
- moderately affected
- unaffected

100% or more of river flow is regulated by dams

Chinese dams

China is the world's most prolific dam builder. Not only is it planning at least 80 large dams and many more smaller ones on its own rivers, but it was estimated in 2008 to be helping to build around 97 dams in 39, mostly developing, countries. Disastrous environmental effects of the Three Gorges Dam has led to China's belated recognition of the need for environmental impact assessments prior to construction, now established international policy.

Euphrates-Tigris

Mae Khlong

Volta

Indian dams

India has 4,300 large dams, submerging an area the size of Switzerland. Hundreds more are planned, including a proposal, currently on hold, to create a "garland" of 37 linked rivers through a series of dams and canals. The world's fifth highest dam, the Tehri, would be part of this scheme. Questions have been raised about siting it in an area of seismic activity. The Tipaimukh Dam in Manipur, which received government approval in 2008, will badly affect the ecology of the Meghna river basin in downstream Bangladesh.

Murray-Darling, Australia

Extensive damming, the over-allocation of the river water by 30%, and years of drought, left some farmers with no irrigation water and the Lower Lakes below sea-level in 2009.

Sedimentation

The build-up of sediment is causing problems in northern and southern Africa, where the reservoirs of at least 30 dams are a quarter full of silt.

China **46%**

USA **14%**

India **9%**

Japan **6%**

Spain **3%**

Canada **2%**

South Korea **2%**

Turkey **1%**

Brazil, France **1%**

South Africa, Mexico, Italy, UK **1%**

Australia **1%**

WORLD DAMS
Percentage of total
2003

**World total:
47,655 dams**

Map copyright 2004 Umeå University and World Resources Institute

9 DRAINING WETLANDS

Wetlands cover

12

million sq km

of the world's surface

Wetlands – bogs, swamps and marshes – play a vital role in the world's water system. Without wetlands, rivers flow too fast, lakes become overburdened with organic matter, and coastlines are eroded.

Only recently has the ecological value of wetlands been recognized. Marshes can detoxify wastewater, and peat bogs are estimated to hold up to a third of the carbon dioxide stored in vegetation and soils. Mangrove swamps play a vital role in protecting the coastline from erosion and storm surges, problems worsening with climate change.

The extent of wetland loss worldwide is unknown, but around half of those in industrialized countries were destroyed during the last century. They used to be seen as unhealthy sources of "swamp fevers" such as malaria, and unproductive "wet deserts" that should be drained for settlement or agriculture. Some have been inundated or dried up as a result of dams altering river flows, others have had their water diverted for irrigation. Despite some reclamation, wetlands are deteriorating faster than any other ecosystem.

Their destruction has a devastating impact on those who live in them and off them. The freshwater fish in inland wetlands provide vital protein for millions of people in developing countries.

Attempts have been made to estimate the economic value of wetlands. In many cases, their worth turns out to be higher than that of the drained land.

In 1971, The Convention on Wetlands was signed at Ramsar in Iran. A list of over 1,800 wetlands of international importance, covering 1.7 million square kilometres, has been drawn up over the intervening years by the 158 signatory states, with the aim of conserving these vital areas. While this international action has helped, it has not been enough to repair damage and stop the rot.

The value of mangroves

Mangroves provide timber and shoreline protection. Their financial value is hard to assess, but ranges from $15,000–$50,000 per square kilometre each year, or as much as $1 million a year in popular tourist areas. Their total value has been estimated at $1.6 billion.

Inner Niger Delta, Mali

During the rainy season, the Niger and Bani rivers overspill their banks to form a 20,000 sq km area of wetland, which supports fishers, farmers, and pastoralists in an otherwise arid country. Growing populations are competing to use the land, while limited rainfall and the damming of the Niger have reduced water levels.

FLORIDA MANGROVES

mangroves

boundary of National Park

Florida Everglades

The 2,000 sq km of Florida's swamps represent only a fraction of their original size. In 1947, their ecological value was recognized by the creation of the Everglades National Park, but revival of the ecosystem has been stalled by the Florida sugar industry and farmers. In 2008, the Florida state government showed its commitment to ecological repair with a landmark decision to buy the US Sugar Corporation for $1.75 billion and use its land to restore flows of water to the Everglades from Lake Okeechobee.

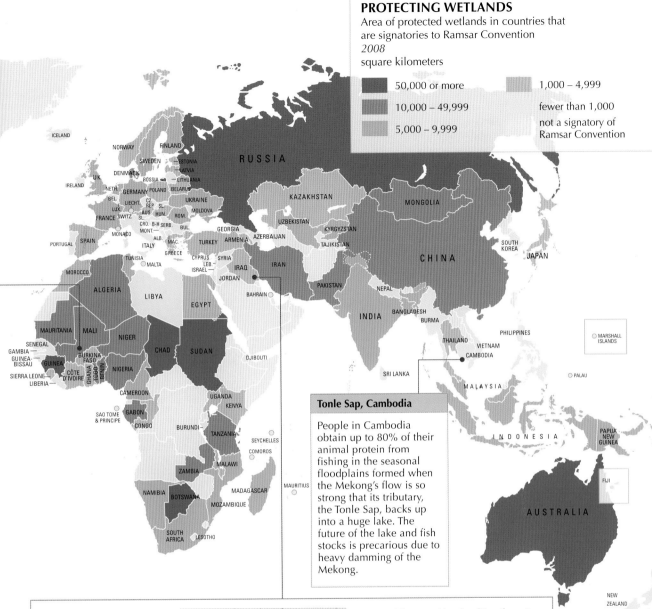

PROTECTING WETLANDS

Area of protected wetlands in countries that are signatories to Ramsar Convention
2008
square kilometers

- 50,000 or more
- 10,000 – 49,999
- 5,000 – 9,999
- 1,000 – 4,999
- fewer than 1,000
- not a signatory of Ramsar Convention

Tonle Sap, Cambodia

People in Cambodia obtain up to 80% of their animal protein from fishing in the seasonal floodplains formed when the Mekong's flow is so strong that its tributary, the Tonle Sap, backs up into a huge lake. The future of the lake and fish stocks is precarious due to heavy damming of the Mekong.

DESTROYING THE GARDEN OF EDEN

Changing extent of the marshlands of southern Iraq

- marshland in 1973
- marshland in 2002
- marshland restored since 2003

The marshlands of Southern Iraq have been almost completely destroyed by dams on the Tigris and Euphrates that substantially reduced their flow and by Saddam Hussein's extensive drainage schemes in the 1990s. Once a richly diverse habitat supporting a population of around 250,000 Ma'dan (Marsh Arabs), they are now largely desert and almost all the Ma'dan have been either killed or displaced. Since the fall of Saddam in 2003, water has been allowed to flow again, and some level of vegetation has spread over half the original area. The way of life of the Ma'dan is not so easily restored.

DRYLANDS AND DROUGHTS

PEOPLE AFFECTED BY DROUGHT

Number in each continent
1999–mid-2008

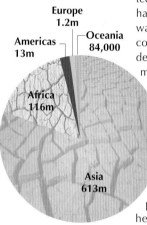

Europe
1.2m

Americas
13m

Oceania
84,000

Africa
116m

Asia
613m

Total number of people affected:
743 million

The area of hyper-arid land increased by

100%

between the 1970s and 2000s

Around 1 billion people live in the world's dryland areas, and their livestock, food and water supplies are particularly vulnerable to droughts and desertification.

Living patterns and food-production techniques, in tropical as in temperate zones, have developed according to the availability of water in the landscape. Many people in dry cold zones such as Mongolia, as well as in hot deserts such as the Sahara, are nomadic and migrate with the seasons in order to exploit scarce water and grazing. But their lifestyles are fragile, and a prolonged multi-year drought can decimate their herds and threaten starvation.

Development of water sources in drylands and deserts – as in Central Asia – can improve farming and livelihood prospects. But the resultant build-up of herds, over-grazing, loss of vegetative cover and topsoil erosion can lead in turn to reductions in rainfall – less moisture in the soil means less moisture in the air – temperature rises, and ultimately to desertification. Poor land and water management, including cutting down trees, is leading to the annual loss of 100,000 square kilometres of arable land, half of which is in Africa – before any consideration is given to the complications of climate change.

Better-watered zones also experience drought when significant reductions in rainfall lead to crop losses and water shortages, although the manipulation of water and trade in food that accompanies industrialization means that lives are not usually lost. Some recent droughts in Europe have put lives at risk from dehydration. Uncontrollable forest fires, as occurred in Greece in 2007, also cause extensive destruction of property and wildlife.

Industrialized societies can flourish in arid conditions, as the development of water-rich lifestyles in desert areas demonstrates. But the exhaustion and degradation of water resources suggests that this is an unsustainable approach to life in drylands.

Spain

Spain's water system depends on an extensive infrastructure, including 1,200 large dams. Several years of drought – 2004/05 was the driest for 70 years – recently revealed the precarious nature of the country's supplies and prompted disputes between regions. In 2008, with reservoirs only 20% full, Catalonia had to ship in water from France.

Las Vegas, USA

Las Vegas, a refrigerated oasis in the Nevada desert, demonstrates the triumph of technology over terrain. But 90% of its current water supply comes from the dwindling Colorado river, and the ever-expanding city will be in water crisis within 15 to 20 years. Plans for a pipeline from another river have met heated opposition.

Mongolia

The lifestyle of Mongolia's pastoralist people is adapted to dry terrain and bitter winters. They knew where to find pastures and water for their herds. But between 1999 and 2002, Mongolia experienced severe drought and extreme temperature – conditions known as *dzud* that normally occur only one year in seven. At least 6 million livestock were lost, and thousands of destitute people migrated to the cities.

ARIDITY ZONES

Land classified according to amount of water lost to atmosphere as a proportion of total rainfall

- hyper-arid
- arid
- semi-arid
- dry sub-humid
- moist sub-humid and humid
- cold

Cyprus

Small islands that rely entirely on rainfall for fresh water are particularly vulnerable to drought. In 2008, reservoirs in Cyprus almost dried out following the worst drought in 100 years, and water had to be shipped in from Greece.

Gobi Desert – on the move

Large parts of China are arid, and the country's food and water supplies are vulnerable to drought. Population pressure and the misuse of land through deforestation and soil erosion have led to the desertification of previously fertile lands. Around 18% of the country is affected, turning tens of millions of people into environmental refugees. The Gobi desert is annually swallowing up nearly 5,000 square kilometres on its way to Beijing.

Australia

The aboriginal people of Australia lived very lightly on the land of this largely dry continent. Post-settlement industrialized farming systems and urban development on the coastal fringes require plentiful supplies of water and are proving hard to sustain. Consecutive years of low rainfall have recently led to crop failures, livestock deaths and extensive wildfires.

Doing battle with the Sahara

The world's largest desert has suffered progressive desiccation over the last 5,000 years. Its small number of plant and animal species, and the 2 million nomadic people dependent on them, are among the most vulnerable in the world. The EU and the African Union have agreed a scheme to create a "green wall of trees" along the edge of the desert, to try and prevent further soil degradation and desertification.

Horn of Africa

The lifestyle of pastoralists in arid regions is adapted to extremely low rainfall. But in 2008, three years of drought, combined with political instability, put 17 million farmers and livestock herders in the Horn of Africa at risk of severe food shortage.

11 FLOODS

FINANCIAL COST

Estimated value of
damage by continent
1999–2008
US$ billions

Life-threatening floods causing widespread destruction to lands, crops and infrastructure are becoming more frequent, and affecting an increasing number of people.

Severe weather triggers floods. Unusually heavy and prolonged rain or rapid seasonal ice-melt, over-fill river basins, spilling streams beyond their banks and floodplains. Hurricanes, cyclones and storms cause the sea to force its way inland, breaching flood defences, as in New Orleans in 2005.

Floods are typically regarded as "natural", meteorologically caused, disasters, whose ferocity and number are increasing as a consequence of climate change. However, man-made alterations of the environment, such as draining of wetlands and constraint of rivers by dams and embankments, are also important factors. Nearly 7 million hectares of wetlands alongside the Mississippi – the river's natural sponge – have been drained and developed over the past 100 years. These areas are now prone to severe flooding in a rainy year.

Deforestation also exacerbates flood disasters. Rainfall on treeless slopes washes away soil that would previously have absorbed it. This increases the loads of both water and silt in rivers and lakes, which reduces the volume of water they can hold.

All riverine and low-lying countries are vulnerable. Some delta areas depend for survival on the annual flood of major rivers and the fertile silt deposited on their soils. One-third of Bangladesh is routinely flooded. If there is too much rain, or a cyclone coincides, the fragile blessings of flood transform into disaster. In 1998, 31 million people in Bangladesh were made homeless and 1,300 lost their lives to rain-induced flooding. Himalayan deforestation and upstream barrages and diversions also contributed.

Floods in cities are made worse by the pervasiveness of impermeable concrete or tarmac, which prevents rain from soaking into the ground. There were flashfloods in Mumbai in 2005 when rainwater could not escape through drains that had become overloaded and blocked with plastic bags.

RISING WATERS
Number of floods worldwide
1997–2007

Guatemala

In October 2005, the torrential rains of Hurricane Stan caused mudslides to engulf dozens of villages, killing 670 people and leaving thousands homeless.

6 Climate Change; 8 Altered Flows; 9 Draining Wetlands

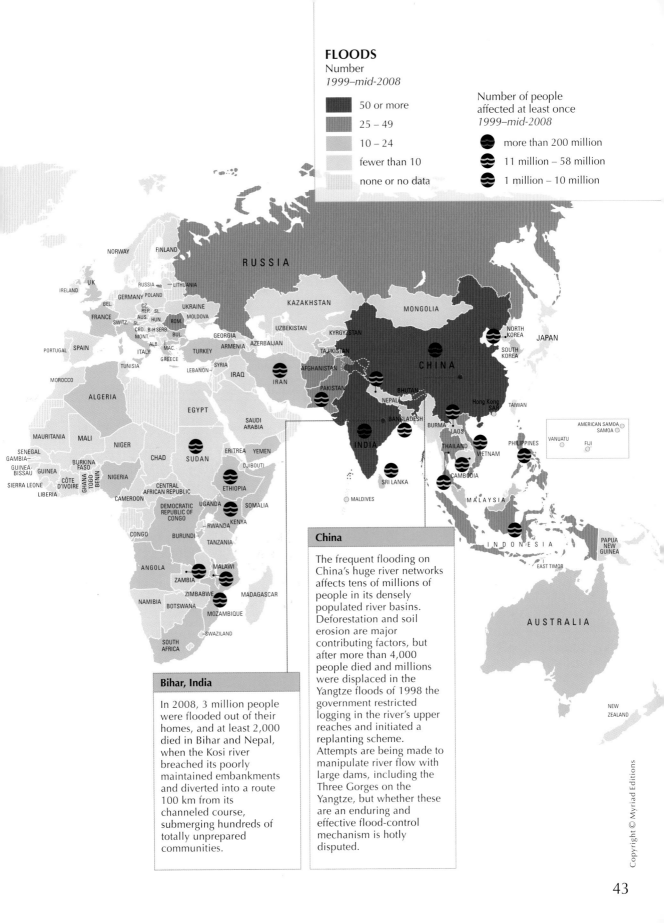

FLOODS

Number
1999–mid-2008

- 50 or more
- 25 – 49
- 10 – 24
- fewer than 10
- none or no data

Number of people
affected at least once
1999–mid-2008

- more than 200 million
- 11 million – 58 million
- 1 million – 10 million

Bihar, India

In 2008, 3 million people were flooded out of their homes, and at least 2,000 died in Bihar and Nepal, when the Kosi river breached its poorly maintained embankments and diverted into a route 100 km from its channeled course, submerging hundreds of totally unprepared communities.

China

The frequent flooding on China's huge river networks affects tens of millions of people in its densely populated river basins. Deforestation and soil erosion are major contributing factors, but after more than 4,000 people died and millions were displaced in the Yangtze floods of 1998 the government restricted logging in the river's upper reaches and initiated a replanting scheme. Attempts are being made to manipulate river flow with large dams, including the Three Gorges on the Yangtze, but whether these are an enduring and effective flood-control mechanism is hotly disputed.

43

PART 3 Water for Living

Drinking water is more vital for human survival than food. Everyone on Earth has access to some source of water, but it is not always safe to consume. Wells and streams that used to be potable have become contaminated with bacteria or other pollutants in today's crowded world. Improved water supplies – piped supplies or covered wells – are highly valued and help control the spread of disease, especially when they are reliable, plentiful and close to home.

However, the amount of water essential for domestic use is far greater than that required purely for drinking. Water is needed for cooking, bathing, washing-up, laundry, flushing the toilet and in some places is also vital for watering small livestock and vegetable plots. Providing adequate piped supplies for every household, especially if treated to drinking-water standards, is extremely expensive, which explains why 1 billion people still have no such service. Many people in developing countries have to manage on much less water than those in industrialized countries, and the poor typically pay more for their supply than the better-off. They are also much less likely to have a tap or flush toilet in the home, let alone a piped connection to a wastewater removal or sewerage system.

Many forms of disease are identified with water, either because the pathogens can be imbibed, or because their hosts or agents – such as the malarial mosquito – breed on standing water. However, the most important connection between water and illness is its use for personal hygiene. Washing hands with soap and preventing contact between human beings and faecal particles by flushing toilets or other forms of sanitation cuts disease transmission dramatically.

Large quantities of water are consumed in the form of food. Edible plants are either nourished by moisture in the soil, known as "green water", or by water channelled from surface or underground sources, known as "blue water". Every calorie we eat requires for cultivation an average of a litre of water. The production of food per head has risen steadily in recent decades, mainly due to intensive planting of high-yielding crop strains under irrigation. But as pressures on food and water supplies increase, new ways of deriving more "crop per drop", especially in rain-fed areas, are needed. The rising price of food reflects in part the rising cost of water.

Water for some people's living has meant its loss for others. More than 80 million people worldwide have been dispossessed of their homes, lands and livelihoods by the construction of large dams since 1950, and others, dependent on fishing or other river-based occupations upstream or downstream, have seen their way of life destroyed. Many received little or no compensation, and the benefits in terms of water for irrigation or hydropower tend to pass them by.

12 WATER FOR DRINKING

884
million
people use an unimproved drinking water source

Everyone has access to a source of drinking water – otherwise they could not survive. But natural sources such as streams and open wells are often contaminated by human and animal waste.

In our increasingly crowded world, access to safe drinking water requires that the supplies people use for domestic purposes reach certain public health standards, known as "improved": they should be close to home, potable, and perennially reliable. This requires a degree of construction or engineering (drilling, laying pipes, building a tank), which means that water – a naturally occurring, free public good – incurs a cost from the "improvement" that someone has to pay for.

Today, 87 percent of people have "access to an improved drinking water source". This is not necessarily at their home, nor does it invariably function properly, but it is within reach. Of those without, the vast majority are rural. However, urban access – reported as 94 percent in developing regions – is often seriously deficient. Townspeople may be assessed as "fully covered" when taps or standpipes are installed every 200 metres, and users are obliged to share with 5,000 others. Village taps or pumps at similar distances serve far fewer people and are far less likely to break down. Furthermore, rapid urbanization means that water supply provision is not keeping pace with population growth in towns and cities.

Those without a tap at home often fetch water from a distance. Traditionally, women haul water on their heads or backs. In many African countries, the round-trip to the source takes more than 30 minutes, restricting the amount that can be brought home. Since this water has to be used for washing, cooking and watering small livestock, as well as for drinking, personal hygiene typically suffers.

WATER ACCESS
Percentage of population using water from different sources
2006

Unimproved
- unprotected well or spring
- water cart
- surface water
- tanker truck

13%

Improved
- public tap or standpipe
- tube well or borehole
- protected dug well
- protected spring
- rainwater collection

33%

- piped water to dwelling or yard

54%

8% Latin America & Caribbean
4% North Africa
9% Sub-Saharan Africa
1% CIS
4% West Asia

CANADA

USA

MEXICO
BAHAMAS
CUBA
DOMINICAN REP.
JAMAICA
HAITI
BELIZE
ANGUILA
ANTIGUA & BARBUDA
GUATEMALA
HONDURAS
ST KITTS & NEVIS
DOMINICA
EL SALVADOR
NICARAGUA
GRENADA
ST LUCIA
BARBADOS
COSTA RICA
TRINIDAD & TOBAGO
PANAMA
VENEZUELA
GUYANA
SURINAME
COLOMBIA
FRENCH GUIANA
ECUADOR

PERU

BRAZIL

BOLIVIA

PARAGUAY

CHILE
ARGENTINA
URUGUAY

WATER COLLECTION
Division of labour by gender
2005–06

boys 4%
men 25%
women
girls 7%

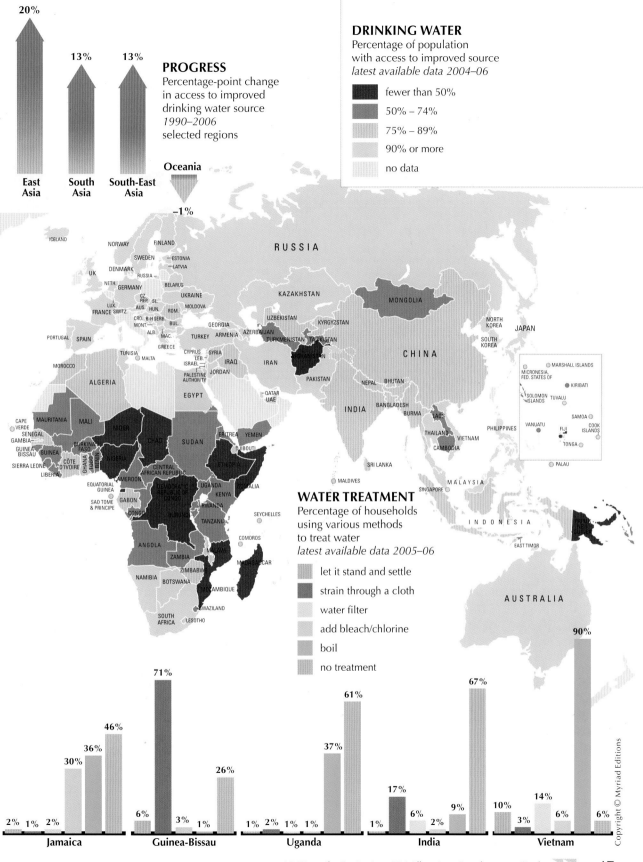

PROGRESS
Percentage-point change
in access to improved
drinking water source
1990–2006
selected regions

20%
East
Asia

13%
South
Asia

13%
South-East
Asia

Oceania
−1%

DRINKING WATER
Percentage of population
with access to improved source
latest available data 2004–06

- fewer than 50%
- 50% – 74%
- 75% – 89%
- 90% or more
- no data

WATER TREATMENT
Percentage of households
using various methods
to treat water
latest available data 2005–06

- let it stand and settle
- strain through a cloth
- water filter
- add bleach/chlorine
- boil
- no treatment

Jamaica
2% 1% 2% 30% 36% 46%

Guinea-Bissau
6% 71% 3% 1% 26%

Uganda
1% 2% 1% 1% 37% 61%

India
1% 17% 6% 2% 9% 67%

Vietnam
10% 3% 14% 6% 90% 6%

Copyright © Myriad Editions

2.5
billion

people have no access to improved sanitation facilities

The most desirable personal convenience flushes into a sewer, but many lower-cost sanitation systems use no water at all.

Conventional sewerage is hugely profligate in water-short areas, typically requiring 15,000 litres per person per year for flushing . The provision of such a service is technically unfeasible in almost all rural and many poor urban environments in developing regions. It would also be prohibitively expensive: costs vary from $400–$1,500 per head, rising steeply if provision is also made for sewage treatment.

The alternative is dry systems, using pits or tanks for storing excreta safe from human contact, and in some cases reusing composted material as fertilizer. These are simpler and cheaper to build, but not to manage. Various models of toilet are regarded as "improved" according to public health definitions, but no distinction is made in coverage data between "wet" and "dry" varieties. On the whole, people adopting toilets in place of the open air prefer a water-seal model, flushed by hand where a sewer connection is impossible.

In some urban environments, a simplified form of sewerage has been developed, using smaller-diameter pipes and community construction and maintenance. This reduces costs to $40–$260 per head. The Orangi Pilot Project in Karachi, Pakistan, has been a pioneer of this approach, and many examples exist in Central and South America.

In developing-world cities, regular sewerage only serves a small elite but its output is often discharged straight into rivers. In India, 80 percent of the pollution destroying the country's rivers is raw sewage. Wastewater services are universally needed – for greywater (water used for cooking, laundry and bathing), blackwater (sewage) and stormwater run-off, which otherwise collects as standing water and breeds insects, including malarial mosquitoes. Given the impossibility of introducing industrialized world wastewater treatment in large parts of the developing world, local management of wastewater is the realistic alternative.

90%

of sewage in developing countries is discharged untreated into watercourses

11% — Latin America & Caribbean

14% — North Africa

5% — Sub-Saharan Africa

CIS — –1%

5% — West Asia

WASTEWATER TREATMENT
Percentage of wastewater treated in selected countries
latest available 1995–2005

Azerbaijan	Botswana	Mexico	Lithuania
3%	19%	19%	26%

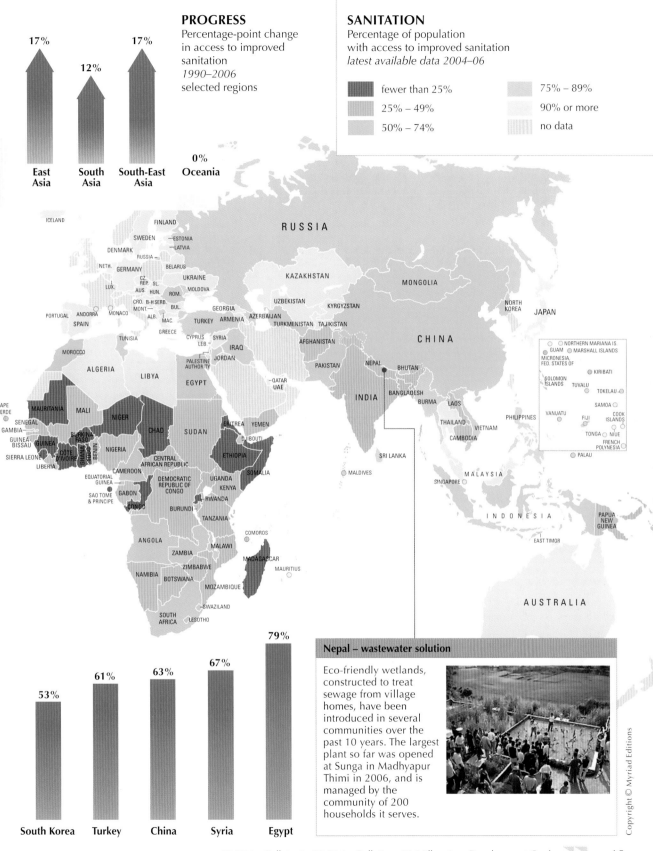

PROGRESS

Percentage-point change
in access to improved
sanitation
1990–2006
selected regions

17% **East Asia**

12% **South Asia**

17% **South-East Asia**

0% **Oceania**

SANITATION

Percentage of population
with access to improved sanitation
latest available data 2004–06

- fewer than 25%
- 25% – 49%
- 50% – 74%
- 75% – 89%
- 90% or more
- no data

53% **South Korea**

61% **Turkey**

63% **China**

67% **Syria**

79% **Egypt**

Nepal – wastewater solution

Eco-friendly wetlands,
constructed to treat
sewage from village
homes, have been
introduced in several
communities over the
past 10 years. The largest
plant so far was opened
at Sunga in Madhyapur
Thimi in 2006, and is
managed by the
community of 200
households it serves.

14 WATER AT HOME

Litres of water used daily in the home:

Australia

282

Ethiopia

13

There are huge discrepancies in the amount of water people use in their home. Much depends on lifestyle and on the availability of water.

In the industrialized world, the amount of water assessed as withdrawn for domestic purposes includes not only that used in the home but that used to wash cars, sprinkle gardens and parks, and fill swimming pools, both private and municipal. In non-industrialized areas, agricultural communities may use domestic water for livestock, and other essential economic activities such as growing vegetables and brewing beer.

Estimates of the amount of water actually used in the home in the industrialized world vary widely, and range from 150 litres to 800 litres a day in hot climates. As environmental awareness has grown, consumers and manufacturers are finding ways of using less water. Efforts to control demand include standard price-control mechanisms: metering and charging higher unit prices. Toilets, showers and washing-machines are nowadays designed to reduce water use.

In poorer regions, where water for the home has to be carried from a pump or tap, or where its purchase represents a major expense, insufficient water may be used for healthy living (definitions of "insufficient" differ, but WHO suggests 20 litres a day). Although the link between water and ill-health is usually associated with drinking-water contamination, a plentiful supply of water via a household tap is actually more important in preventing transmission of disease. A safe drinking water supply reduces diarrhoeal disease by 6 percent; while improved hygiene, especially washing hands with soap, can reduce diarrhoeal incidence by 45 percent.

Initiatives to ensure a cheap and accessible supply of water to people's homes are therefore vital, as are those that build awareness about using water for hygiene in disease-prone environments. Plans for household water services also need to allow for sufficient supplies for a wide variety of purposes related to family survival, depending on the local economy and culture.

12
power shower
6
water-saver
shower head

Hand-washing priorities

Education about hand washing needs to focus on what makes the greatest difference. Researchers in Guatemala found that women who washed before or after contact with food, water or their children, and after going to the toilet or changing a baby's nappy used an extra 20 litres of water, taking up an hour each day. In fact, it is on the last two occasions when hand washing is most important.

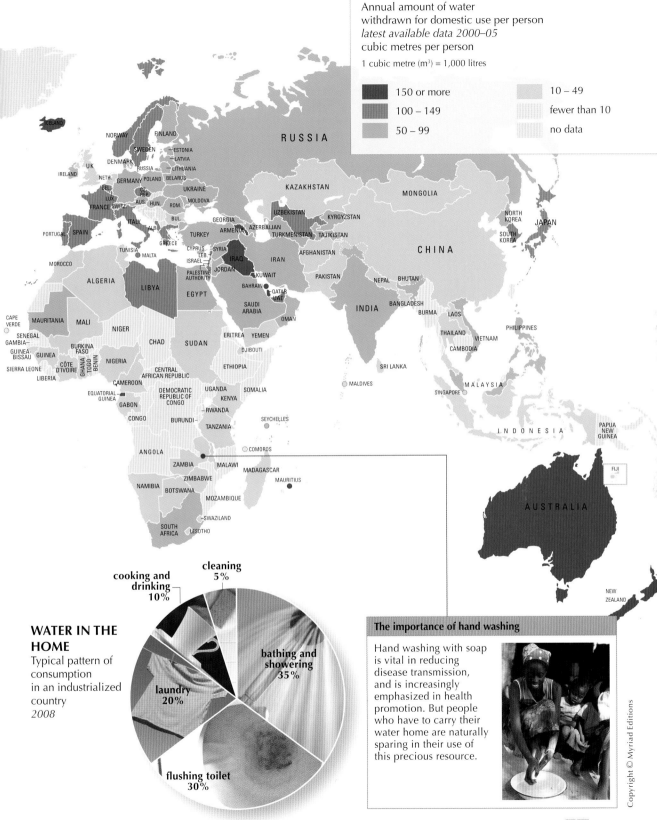

DOMESTIC WATER USE

Annual amount of water
withdrawn for domestic use per person
latest available data 2000–05
cubic metres per person

1 cubic metre (m³) = 1,000 litres

- 150 or more
- 100 – 149
- 50 – 99
- 10 – 49
- fewer than 10
- no data

WATER IN THE HOME

Typical pattern of
consumption
in an industrialized
country
2008

cleaning
5%

cooking and
drinking
10%

bathing and
showering
35%

laundry
20%

flushing toilet
30%

The importance of hand washing

Hand washing with soap
is vital in reducing
disease transmission,
and is increasingly
emphasized in health
promotion. But people
who have to carry their
water home are naturally
sparing in their use of
this precious resource.

Copyright © Myriad Editions

17 Water for Food; 24 Water for Sale

15 WATER AND DISEASE

An estimated

10%

of the global disease burden could be prevented by improving water supplies, sanitation and hygiene behaviour

Water and disease interact in two ways: unsafe drinking water can spread disease, but water used for personal and domestic hygiene can prevent disease transmission.

The real source of much disease labelled as "water-related" is lack of sanitation (on-site or dry systems as well as waterborne sewerage), and low standards of hygiene – itself heavily influenced by the quantity of water in the home. Where faeces are not safely confined from human contact, especially when deposited in the open or in waterways, feet, hands or mouths come into contact with the bacteria, viruses and parasites they contain. Via a variety of pathways, or "faecal–oral" routes of transmission, these end up in people's digestive tracts.

The diarrhoeal diseases – notably cholera, typhoid and dysentery – are classic examples of infections transmitted by faecal-oral routes. Ensuring that drinking water is safe, both at the source and in containers stored at home is an important, but only a partial, means of protection. Cutting off other faecal-oral routes via sanitation and personal hygiene, especially hand-washing, needs much more attention in strategies for protecting health.

Poor hygiene plays a role in the transmission of almost every disease. Any infection in the body – especially intestinal worms or helminths, mostly contracted from the soil – lowers nutritional status, thus connecting diseases of poor hygiene with childhood malnutrition. Certain conditions are directly associated with failure to wash, and therefore with acute water-short lifestyles, such as those of people living in severely deprived or arid areas, or both. Trachoma, which impairs sight, and scabies, a skin disease, are both in this category.

Diseases related to lack of water, sanitation and hygiene not only kill but affect people's ability to work and look after themselves and their families. They are among the major causes of poverty.

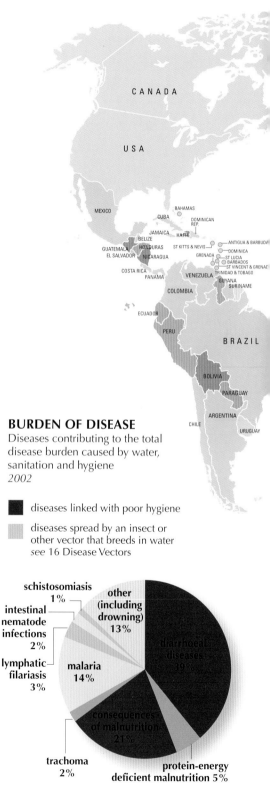

BURDEN OF DISEASE
Diseases contributing to the total disease burden caused by water, sanitation and hygiene
2002

- ■ diseases linked with poor hygiene
- ▨ diseases spread by an insect or other vector that breeds in water
  see 16 Disease Vectors

schistosomiasis 1%
intestinal nematode infections 2%
lymphatic filariasis 3%
trachoma 2%
protein-energy deficient malnutrition 5%
consequences of malnutrition 21%
malaria 14%
other (including drowning) 13%
diarrhoeal diseases 39%

◄◄ *7 Urbanization; 12 Water for Drinking; 13 Water for Sanitation; 14 Water at Home*

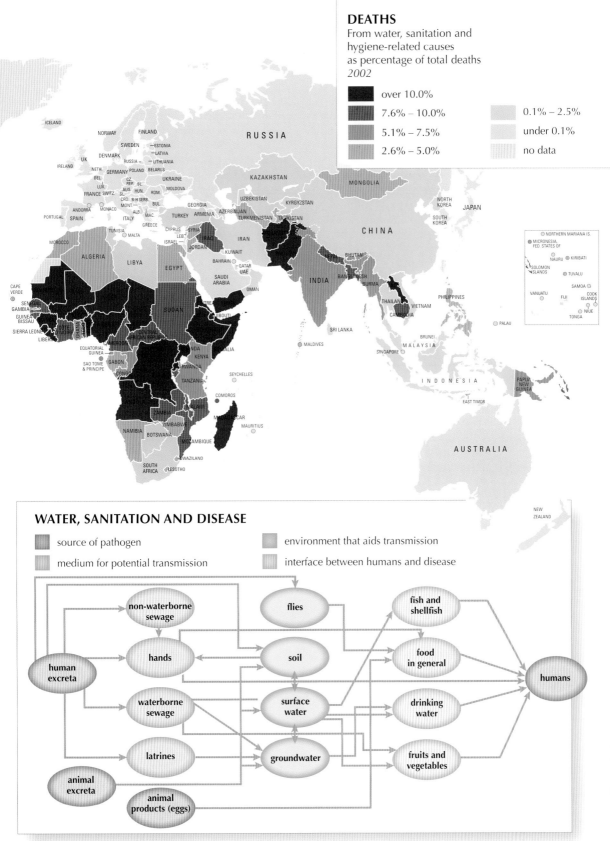

DEATHS

From water, sanitation and hygiene-related causes as percentage of total deaths
2002

- over 10.0%
- 7.6% – 10.0%
- 5.1% – 7.5%
- 2.6% – 5.0%
- 0.1% – 2.5%
- under 0.1%
- no data

WATER, SANITATION AND DISEASE

- source of pathogen
- medium for potential transmission
- environment that aids transmission
- interface between humans and disease

human excreta • non-waterborne sewage • hands • waterborne sewage • latrines • animal excreta • animal products (eggs) • flies • soil • surface water • groundwater • fish and shellfish • food in general • drinking water • fruits and vegetables • humans

16 DISEASE VECTORS

**A child
dies of malaria
every**

30

seconds

Water is the breeding-ground for many vectors of disease, including flying insects such as malarial mosquitoes, snails and parasitic worms.

Malaria is the best-known and most deadly of these diseases, causing nearly 1 million deaths a year and placing a huge burden on affected families and medical services. Transmission occurs when certain species of mosquito (*Anopheles* is the most dangerous) spread malarial parasites from one victim's bloodstream to that of others via the mosquito's feeding method (blood-sucking). Dengue fever, lymphatic filariasis (painful swelling of the limbs), and West Nile Fever – now found in the USA – are spread in the same way.

Other water-breeding flying insects such as blackflies and tsetse flies transmit diseases such as river blindness (onchocerciasis) and African sleeping sickness – conditions known since ancient times and once widely dispersed, but now confined to certain tropical, humid, and deeply impoverished areas. Around 37 million people are infected with onchocerciasis, which causes severe skin disease, visual impairment, and blindness. It can shorten life expectancy by up to 15 years.

Another type of infection is contracted from wading in surface water polluted with human faeces. These infections include schistosomiasis (or bilharzia), the parasite for which inhabits a water-snail living in still or stagnant water. More than 200 million people suffer from the

disease, which can cause abdominal pain and enlargement of the liver. It results in 200,000 deaths a year.

New reservoirs behind dams may expand the breeding-grounds both for snails and flying insects – as does any increased presence of standing water. However, in the case of some water-related projects, increased prosperity, improved nutrition and access to medical facilities appear to outweigh the additional risk of infection.

The guinea-worm parasite also breeds in water, and people can become infected when they drink from an infected source. The larvae mature into long worms inside the human body which slowly emerge through the skin, often in the arm or leg. Bathing the painful, incapacitating sore with the protruding worm in an open water source re-contaminates the water and continues the cycle. Efforts to eradicate it have resulted in a dramatic decline in the number of people affected from over 3.5 million in the 1980s to 10,000 in 2005, and it is now endemic only in nine countries in Sub-Saharan Africa.

Many of these water-borne diseases continue to exist because of a lack of hygiene, poor drainage and ignorance. Several were designated Neglected Tropical Diseases by WHO in 2007, and new efforts are being made to eradicate them.

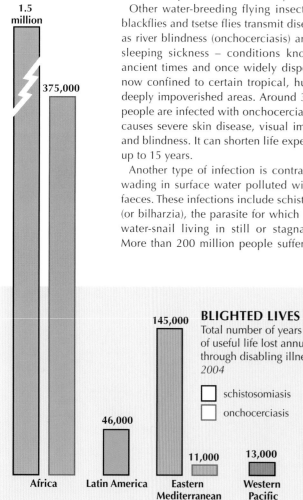

1.5 million

375,000

145,000

BLIGHTED LIVES
Total number of years
of useful life lost annually
through disabling illness
2004

- ☐ schistosomiasis
- ☐ onchocerciasis

46,000

11,000

13,000

Africa   Latin America   Eastern Mediterranean   Western Pacific

MALARIA DEATHS
Number by WHO region
2004

Malaria affects 500 million people worldwide. Poverty, inadequate water resources and underfunded health services make it difficult to eradicate. The disease, in turn, places a huge financial burden on the poor. It is estimated that each bout of malaria costs 10 working days, and that malaria accounts for up to 30 percent of hospital admissions in tropical Africa.

YEARS OF DISABILITY
Total number of years of useful life lost annually through disabling illness transmitted by mosquitoes
2004

☐ lymphatic filariasis

☐ dengue

Over 120 million people in 80 countries are thought to be infected with **lymphatic filariasis**, which causes chronic swelling and recurrent secondary bacterial infections. It has a major social and economic impact on their lives.

Dengue causes fever, rash, muscle and joint pain. It is on the increase, with urbanization bringing increased opportunities for the spread of infection and for the breeding of the mosquito vector, *Aedes aegypti*. Up to 50 million infections occur each year, with 500,000 turning into the more severe Dengue Haemorrhagic fever (DHF), which causes 22,000 deaths, mainly of children.

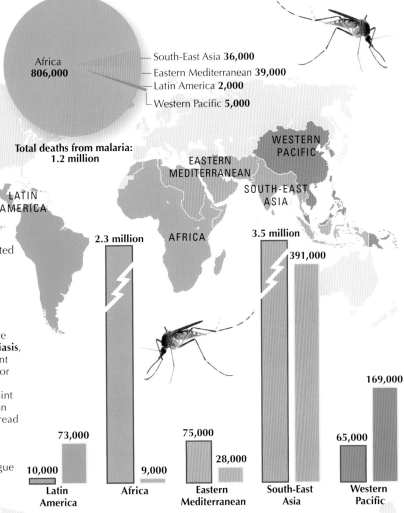

Africa **806,000**

South-East Asia **36,000**
Eastern Mediterranean **39,000**
Latin America **2,000**
Western Pacific **5,000**

Total deaths from malaria: 1.2 million

WESTERN PACIFIC

EASTERN MEDITERRANEAN

SOUTH-EAST ASIA

LATIN AMERICA

AFRICA

	2.3 million		3.5 million	
			391,000	
				169,000
73,000		75,000		
10,000	9,000	28,000	65,000	
Latin America	**Africa**	**Eastern Mediterranean**	**South-East Asia**	**Western Pacific**

WEST NILE VIRUS IN USA
Number of cases
2007

☐ 300 or more
☐ 100 – 299
☐ 50 – 99
☐ fewer than 50
☐ none

West Nile virus was first identified in 1937 in Uganda, and is found in tropical Africa and in India. It is carried in birds and mammals, and transmitted to humans by mosquitoes, where it can develop into encephalitis. In 1999, it was first detected in the USA, where it appears to have developed new transmission routes, including blood transfusions and organ donations. Since 2003, all donated blood in the USA is screened for the virus, and public health campaigns have alerted people to the risks associated with mosquito bites.

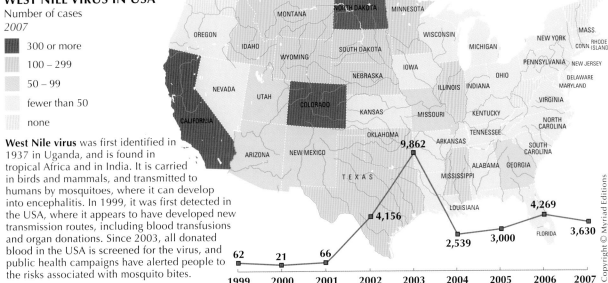

1999	2000	2001	2002	2003	2004	2005	2006	2007
62	21	66	4,156	9,862	2,539	3,000	4,269	3,630

CHANGING DIET IN CHINA

Weight of meat and milk purchased annually by urban dwellers
1990–2006

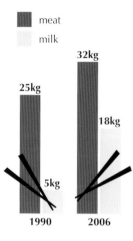

■ meat
░ milk

1990: 25kg, 5kg
2006: 32kg, 18kg

All food production depends on water. Food crops flourish thanks to the rain or hydraulic manipulation.

Every calorie we eat requires for its production roughly one litre of water. So, a good daily diet requires around 3,000 litres, compared to 2 to 5 litres for drinking. Food cultivation therefore demands huge water resources, although the amount required per kilogramme varies widely, with meat from cattle fed on grain requiring the most.

Around 80 percent of agricultural water use is from rainfall stored in the soil, known as "green water", with the rest from "blue water" – rivers, lakes and aquifers. Blue-water withdrawals grew rapidly during the past century.

Originally, agriculture was totally controlled by climate and environment. The fertility provided by rivers and wetlands carrying extra water in the wet season had to be carefully exploited, especially in rain-short areas. Crops were – and are – grown on floodplains when waters recede, streams channelled for irrigation, and rainwater harvested by storing it behind small dams. Growing food has always been about creative manipulation of water – and imprecations to the skies.

Capturing temporary flows is vital in regions with minimal rain (Australia, Egypt and some US states), or where it arrives annually in short bursts (Asia). These regions depend heavily on blue water for farming, and pioneered the construction of massive irrigation projects to expand crop production – the original golden bullet for increasing food supplies.

Food production and consumption per head has risen steadily since the 1960s. This is largely due to the intensive planting of high-yielding crop strains, mostly under irrigation, in Asia in particular. But productivity in rain-fed areas unsuited to large-scale irrigation is lagging. Meanwhile, in rapidly developing economies, foods requiring more water to produce, such as meat, are in greater demand. Industrialization also competes with agriculture for engineered freshwater resources, building pressure for "more crop per drop".

The relationship between food and water shortage is complicated by inequalities of wealth and power, and by climate vulnerability. There are 850 million undernourished people in today's world. Although many live in arid areas, others live in water-rich but densely populated river basins, such as Bangladesh, a hunger hot-spot. Enabling families to grow or buy more food requires efficient water management and the promotion of crops requiring less water.

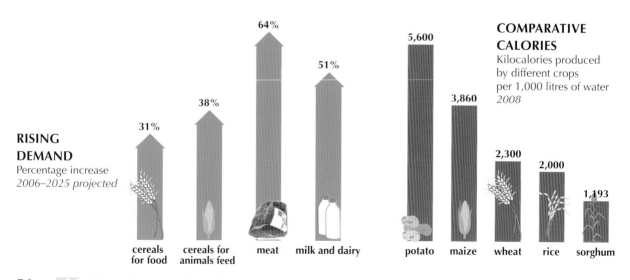

RISING DEMAND
Percentage increase
2006–2025 projected

- cereals for food: 31%
- cereals for animals feed: 38%
- meat: 64%
- milk and dairy: 51%

COMPARATIVE CALORIES
Kilocalories produced by different crops per 1,000 litres of water
2008

- potato: 5,600
- maize: 3,860
- wheat: 2,300
- rice: 2,000
- sorghum: 1,193

3 Rising Demand; 6 Climate Change; 10 Drylands and Droughts

WATER FOR AGRICULTURE

Amount withdrawn for agricultural use
2003 or latest available
cubic metres per person per year

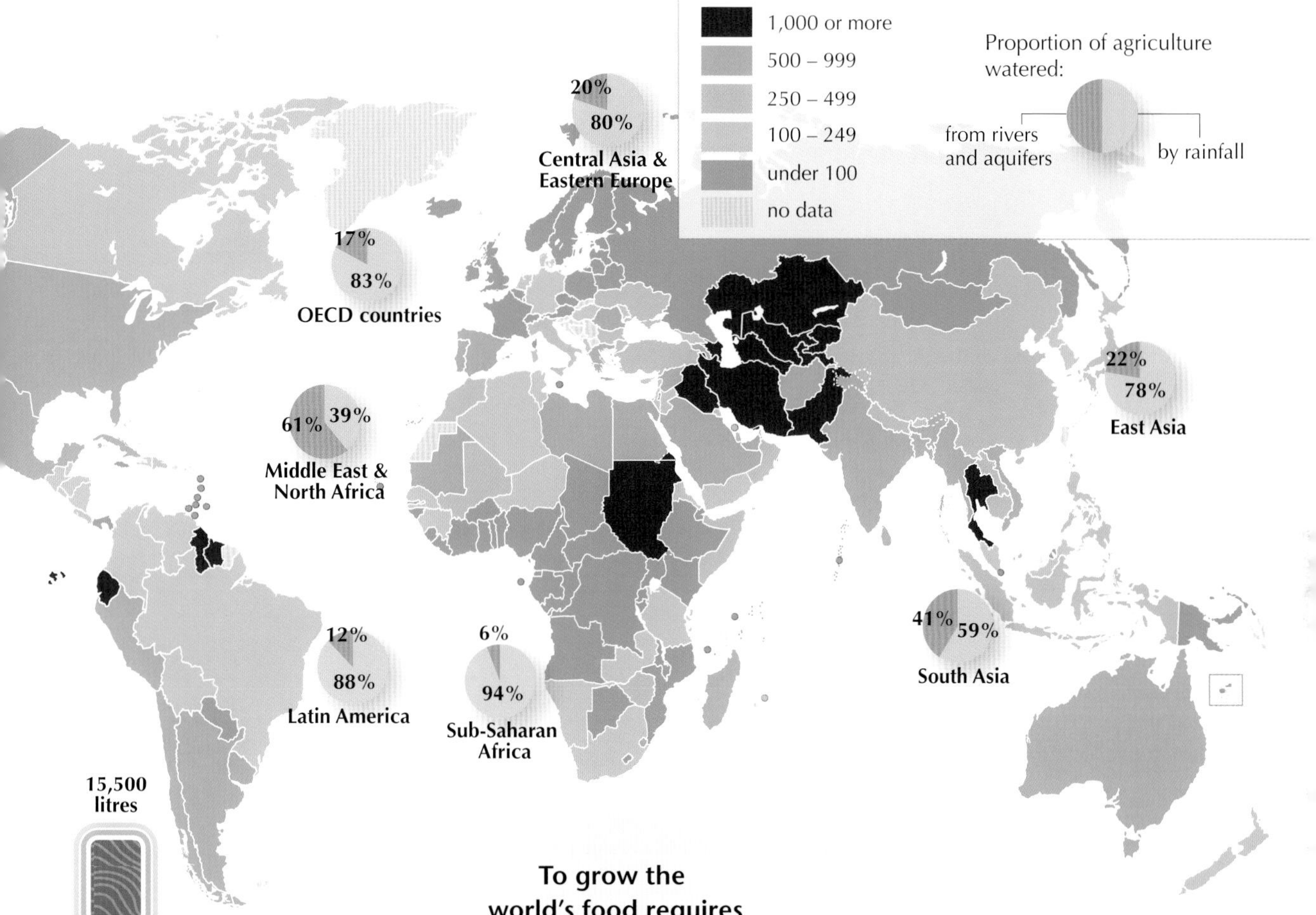

- 1,000 or more
- 500 – 999
- 250 – 499
- 100 – 249
- under 100
- no data

Proportion of agriculture watered:

from rivers and aquifers — by rainfall

20%
80%
Central Asia &
Eastern Europe

17%
83%
OECD countries

22%
78%
East Asia

61% 39%
Middle East &
North Africa

12%
88%
Latin America

6%
94%
Sub-Saharan
Africa

41% 59%
South Asia

To grow the
world's food requires

90

times the annual flow
of the Nile

**15,500
litres**

**5,000
litres**

**5,000
litres**

**4,043
litres**

**3,918
litres**

**2,975
litres**

**2,853
litres**

**1,300
litres**

**625
litres**

beef | cheese | millet | goat | poultry | rice (husked) | sorghum | wheat | potato

WATER FOR FOOD

Average amount of water
needed to produce
1 kilogram of food
2004

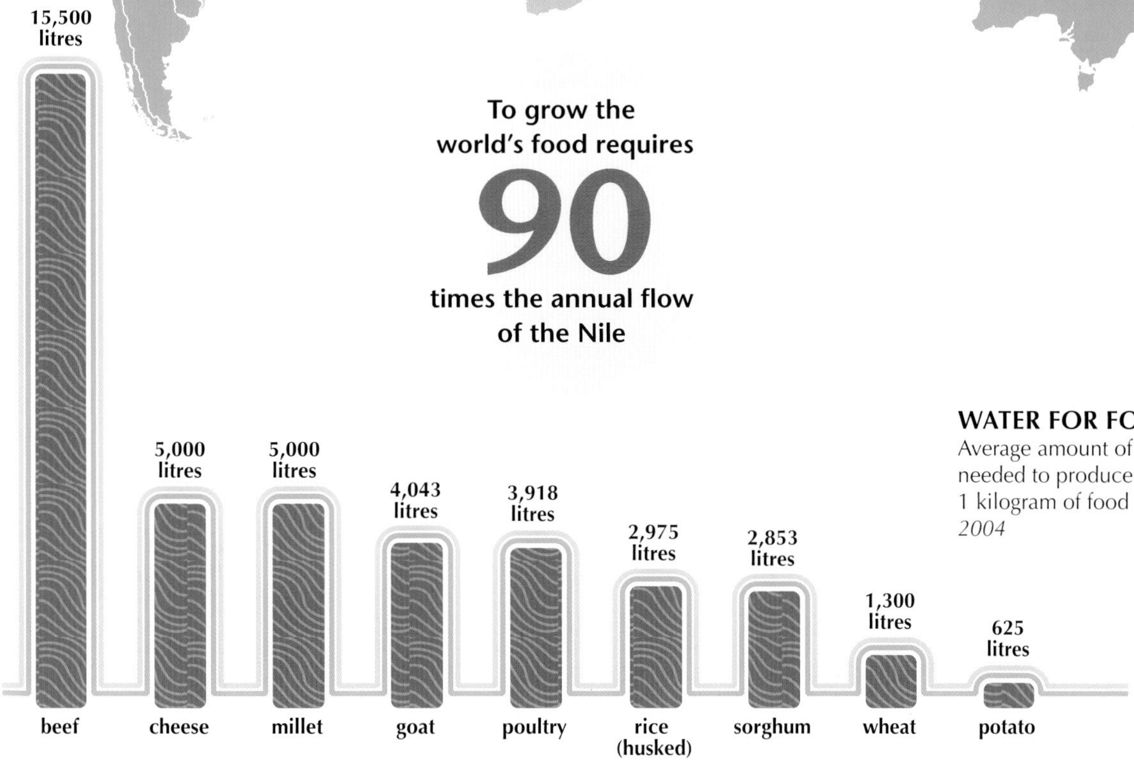

18 DISPOSSESSION BY WATER

By 2020, the Three Gorges dams will have involved relocating

5
million
people

Dams and disease

Globally, 18.3 million people are at risk from malaria and 42 million from schistosomiasis simply because they live near dams, like these children, who live close to the Nam Leuk dam in Laos. Slow-moving water in reservoirs or irrigation channels allows disease-bearing organisms to flourish. The Manantali and Diama dams on the Senegal river led to a dramatic increase in malaria and schistosomiasis, claiming up to 8,000 deaths a year. The Nam Leuk dam in Thailand, completed in 1996, also led to increases in snails and schistosome parasites on the shores of the lake behind the dam.

"Water for living", means not just the water necessary for eating, drinking and health, but involves managing rivers to sustain the lives of those dependent on them.

The corralling of water behind large dams, whatever the potential economic benefits, has many negative impacts on basin inhabitants. Around 80 million people have been dispossessed of their homes, lands and livelihoods by large dams since 1950. Most received inadequate or no compensation, and a disproportionate number belong to ethnic minorities or indigenous peoples whose rights are flouted. Many millions more have had their livelihoods or health threatened by the continuing impact of river impoundment, and rarely enjoy any benefits, such as electricity.

China is the largest dam builder in the world. Many dams under construction are in densely populated areas, requiring the relocation of millions of people, and some are in areas subject to seismic activity. Large dams in India have also caused the forcible relocation of millions, many of them tribals or *adivasi*, and generated huge protests. Corruption associated with these massive projects often extends to the compensation allocation, which vanishes into the wrong hands.

Although those relocated are often promised "land for land", sufficient land is rarely available, so they are given cash instead. Once that is exhausted, they end up on the scrapheap of the urban poor. Those not dispossessed may find their livelihoods ruined by the ecological makeover wrought by the dam. Fisheries and forests are permanently affected, and traditional rights to exploit these natural resources lost.

Pollution that cannot be dispersed by an impounded stream also affects riverine communities, as do sudden floods caused by dam breaches or emergency floodwater releases. Large reservoirs can also be breeding-grounds for disease-carrying mosquitoes and water-dwelling parasites.

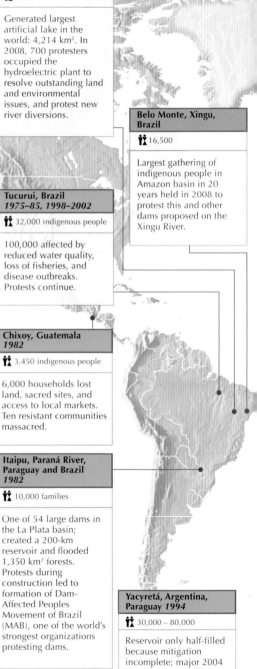

Sobradinho, Bahia, Brazil, 1978
70,000

Generated largest artificial lake in the world: 4,214 km². In 2008, 700 protesters occupied the hydroelectric plant to resolve outstanding land and environmental issues, and protest new river diversions.

Belo Monte, Xingu, Brazil
16,500

Largest gathering of indigenous people in Amazon basin in 20 years held in 2008 to protest this and other dams proposed on the Xingu River.

Tucurui, Brazil 1975–85, 1998–2002
32,000 indigenous people

100,000 affected by reduced water quality, loss of fisheries, and disease outbreaks. Protests continue.

Chixoy, Guatemala 1982
3,450 indigenous people

6,000 households lost land, sacred sites, and access to local markets. Ten resistant communities massacred.

Itaipu, Paraná River, Paraguay and Brazil 1982
10,000 families

One of 54 large dams in the La Plata basin; created a 200-km reservoir and flooded 1,350 km² forests. Protests during construction led to formation of Dam-Affected Peoples Movement of Brazil (MAB), one of the world's strongest organizations protesting dams.

Yacyretá, Argentina, Paraguay 1994
30,000 – 80,000

Reservoir only half-filled because mitigation incomplete; major 2004 enquiry prompted by 4,000 affected families.

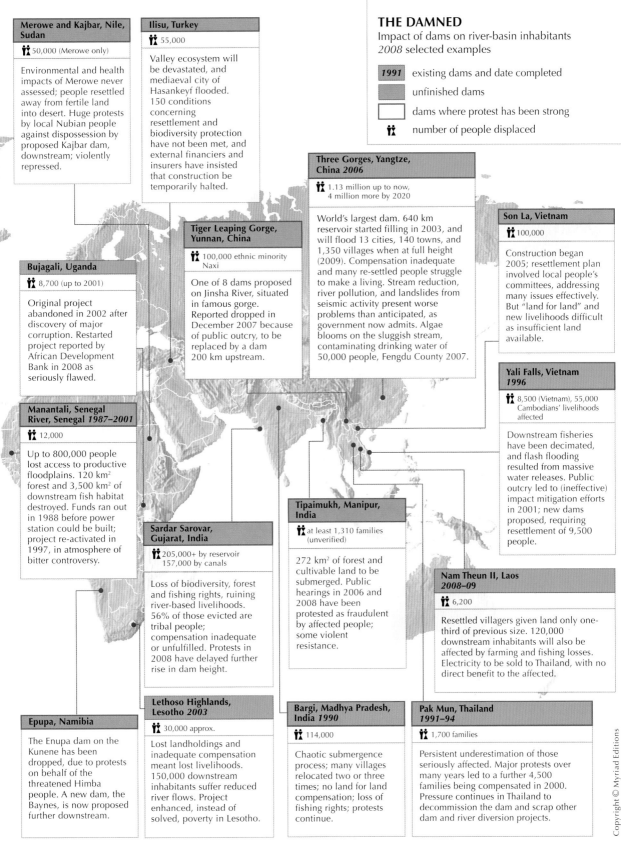

THE DAMNED

Impact of dams on river-basin inhabitants
2008 selected examples

1991 existing dams and date completed

unfinished dams

dams where protest has been strong

number of people displaced

Merowe and Kajbar, Nile, Sudan

50,000 (Merowe only)

Environmental and health impacts of Merowe never assessed; people resettled away from fertile land into desert. Huge protests by local Nubian people against dispossession by proposed Kajbar dam, downstream; violently repressed.

Ilisu, Turkey

55,000

Valley ecosystem will be devastated, and mediaeval city of Hasankeyf flooded. 150 conditions concerning resettlement and biodiversity protection have not been met, and external financiers and insurers have insisted that construction be temporarily halted.

Three Gorges, Yangtze, China 2006

1.13 million up to now, 4 million more by 2020

World's largest dam. 640 km reservoir started filling in 2003, and will flood 13 cities, 140 towns, and 1,350 villages when at full height (2009). Compensation inadequate and many re-settled people struggle to make a living. Stream reduction, river pollution, and landslides from seismic activity present worse problems than anticipated, as government now admits. Algae blooms on the sluggish stream, contaminating drinking water of 50,000 people, Fengdu County 2007.

Son La, Vietnam

100,000

Construction began 2005; resettlement plan involved local people's committees, addressing many issues effectively. But "land for land" and new livelihoods difficult as insufficient land available.

Tiger Leaping Gorge, Yunnan, China

100,000 ethnic minority Naxi

One of 8 dams proposed on Jinsha River, situated in famous gorge. Reported dropped in December 2007 because of public outcry, to be replaced by a dam 200 km upstream.

Bujagali, Uganda

8,700 (up to 2001)

Original project abandoned in 2002 after discovery of major corruption. Restarted project reported by African Development Bank in 2008 as seriously flawed.

Yali Falls, Vietnam 1996

8,500 (Vietnam), 55,000 Cambodians' livelihoods affected

Downstream fisheries have been decimated, and flash flooding resulted from massive water releases. Public outcry led to (ineffective) impact mitigation efforts in 2001; new dams proposed, requiring resettlement of 9,500 people.

Manantali, Senegal River, Senegal 1987–2001

12,000

Up to 800,000 people lost access to productive floodplains. 120 km² forest and 3,500 km² of downstream fish habitat destroyed. Funds ran out in 1988 before power station could be built; project re-activated in 1997, in atmosphere of bitter controversy.

Tipaimukh, Manipur, India

at least 1,310 families (unverified)

272 km² of forest and cultivable land to be submerged. Public hearings in 2006 and 2008 have been protested as fraudulent by affected people; some violent resistance.

Nam Theun II, Laos 2008–09

6,200

Resettled villagers given land only one-third of previous size. 120,000 downstream inhabitants will also be affected by farming and fishing losses. Electricity to be sold to Thailand, with no direct benefit to the affected.

Sardar Sarovar, Gujarat, India

205,000+ by reservoir 157,000 by canals

Loss of biodiversity, forest and fishing rights, ruining river-based livelihoods. 56% of those evicted are tribal people; compensation inadequate or unfulfilled. Protests in 2008 have delayed further rise in dam height.

Epupa, Namibia

The Enupa dam on the Kunene has been dropped, due to protests on behalf of the threatened Himba people. A new dam, the Baynes, is now proposed further downstream.

Lethoso Highlands, Lesotho 2003

30,000 approx.

Lost landholdings and inadequate compensation meant lost livelihoods. 150,000 downstream inhabitants suffer reduced river flows. Project enhanced, instead of solved, poverty in Lesotho.

Bargi, Madhya Pradesh, India 1990

114,000

Chaotic submergence process; many villages relocated two or three times; no land for land compensation; loss of fishing rights; protests continue.

Pak Mun, Thailand 1991–94

1,700 families

Persistent underestimation of those seriously affected. Major protests over many years led to a further 4,500 families being compensated in 2000. Pressure continues in Thailand to decommission the dam and scrap other dam and river diversion projects.

PART 4 Water for Economic Production

Water plays a central part in all economic productivity, either directly as an input, or as part of the process or context in which economic activity takes place. The tourist and recreation industries, too, would be at a loss without water not only for swimming pools and golf courses but also in the natural environment to visit, swim in, cruise on, or just enjoy.

Water is essential for agriculture. Four-fifths of the world's cropland can manage with rain, but the rest needs irrigation from rivers or other "blue water" supplies, either to bring in extra crops per year or to make the desert bloom. The burst of industrial-scale impoundments and irrigation networks that took place during the second half of the 20th century has now slowed. Small-scale irrigation, which is less expensive and environmentally disruptive, now commands more attention, as does more efficient use of moisture in the soil and expansion of "green water" yields.

Agriculture may still absorb the lion's share of water for economic production – accounting for 70 percent of extractions globally – but industrial usage is rising. In high-income countries, the proportion already amounts to 60 percent. Certain industries use more water than others, the paper and metal industries in particular. The energy industry also requires water, for powering turbines to generate electricity, as a cooling agent in power stations, and to grow substitute bio-fuels. Concerns over water use in industry tend to focus less on the volume extracted, however, than on uncontrolled discharges of polluted effluent.

Aquaculture, or fish farming, has rapidly expanded in recent years as wild freshwater fish stocks have equally rapidly declined. Migratory fish and other organisms essential to support the aquatic ecosystem are not able to pass dams, and the fragmentation of rivers has therefore seriously affected traditional fishing livelihoods. However, the growth of commercial fish farming, especially in Asia, has opened up new employment opportunities as well as contributing to the global food system. Most of this produce is exported to industrialized countries to satisfy their demand for fish and seafood. But lack of regulation and the degradation of coastal mangroves to make way for fish farms have imposed considerable environmental costs.

Economic activity around our consumption of water for drinking and domestic use has also increased in recent years. The privatization of water utilities has focused attention on the rising price of domestic supplies, especially for the poor, and whether profits should be made out of trade in such an essential life-giving substance remains controversial. Meanwhile, the bottled water industry has hugely expanded. Once again, the environmental costs of the industry are heavy, and questions as to whether the end product is really superior to water from the tap are beginning to fuel a backlash.

19 IRRIGATION

UNEQUAL DEVELOPMENT

Percentage of appropriate land equipped for irrigation
1999

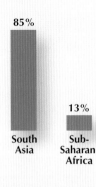

85%
South Asia

13%
Sub-Saharan Africa

Some methods of irrigation, such as spraying, can be very wasteful of water, unlike drip irrigation (shown here), in which water leaks from hoses at the base of the crops.

While such targeted water leakage is very effective, irrigation systems worldwide have a tendency to leak water unintentionally, with as much as half the water withdrawn never reaching the crop.

A fifth of the world's cropland cannot support agriculture purely on rainfall ("green water"), so farmers use water from rivers, lakes, or aquifers ("blue water") to cultivate their crops.

In water-short and monsoon environments, hydraulic systems of irrigation have been used for centuries. Industrial-scale irrigation took off in the 20th century, accelerating in the 1960s with the advent of modern agricultural techniques, which involve high-yielding plant varieties being combined with fertilizers and pest control, and rely on irrigation to promote optimum growth.

The area of irrigated land has increased significantly since then, although it still amounts to only 28 percent of harvested land. It makes a substantial contribution to the global food basket, with irrigated land producing 60 percent of cereals in developing countries – an important source of calories.

Dams provide around 30 percent of irrigation water, although the massive constructions and networks of irrigation canals so popular in Asia have technical drawbacks beyond the social and environmental damage they cause. Much irrigated land has become waterlogged and 10 percent has been lost to salinity. Africa has seen relatively few of these large projects; technical obstacles make them extremely costly per irrigated hectare.

The majority of irrigation water is pumped up from aquifers, and this blue water source is increasingly under threat. In India, groundwater is being depleted far faster than it can be recharged. Over 20 million small electric pumps are in use, affordable even by modest farming families thanks to cheap electricity. As a result, shallower aquifers have dried up, river flows have been reduced, and the water table is dropping out of reach.

With expanded irrigation no longer a magic bullet for increasing food production, attention is returning to green water. Ways are being sought of capturing and storing rainfall, and improving soil quality so that it holds moisture for longer.

Around

70%

of water withdrawals are for irrigation

Brazil **0.5m**

Syria **0.6m**

Turkey **0.7m**

Italy **0.9m**

Saudi Arabia **1.5m**

Mexico **1.7m**

Bangladesh **2.6m**

Iran **3.6m**

Pakistan **4.9m**

China **8.8m**

USA **10.8m**

India **26.5m**

GROUNDWATER
Area under irrigation by groundwater
2007
hectares

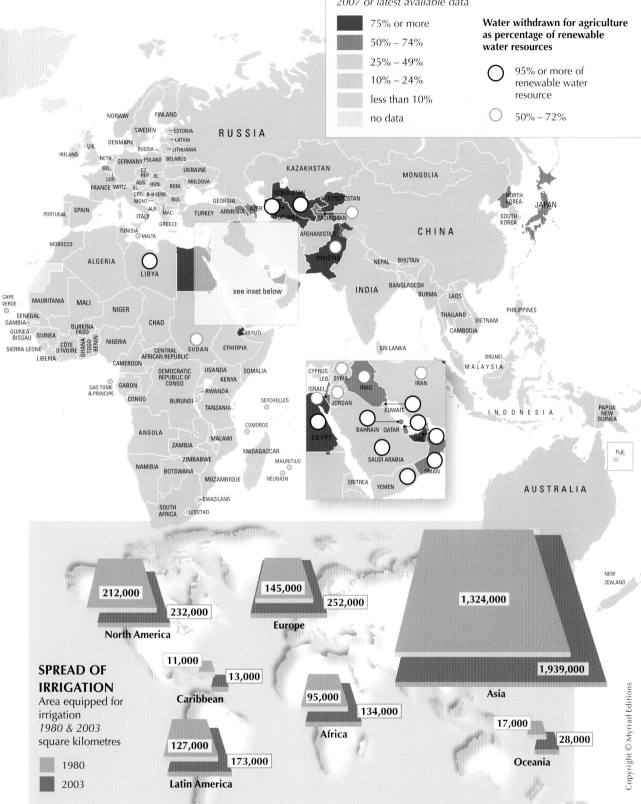

IRRIGATED LAND

As percentage of cultivated area
2007 or latest available data

- 75% or more
- 50% – 74%
- 25% – 49%
- 10% – 24%
- less than 10%
- no data

Water withdrawn for agriculture as percentage of renewable water resources

- 95% or more of renewable water resource
- 50% – 72%

see inset below

SPREAD OF IRRIGATION

Area equipped for irrigation
1980 & 2003
square kilometres

- 1980
- 2003

North America 212,000 / 232,000

Europe 145,000 / 252,000

Asia 1,324,000 / 1,939,000

Caribbean 11,000 / 13,000

Africa 95,000 / 134,000

Latin America 127,000 / 173,000

Oceania 17,000 / 28,000

Copyright © Myriad Editions

20 WATER FOR INDUSTRY

Water used for industrial purposes:

776
cubic kilometres a year

WATER FOR BIO-FUELS

Litres of water needed to grow and process crops to produce 1 litre of fuel
2008

bio-ethanol   bio-diesel

up to **10,000**

up to **20,000**

Just over 20 percent of all fresh water withdrawn worldwide is used by industry, although in high-income countries this proportion rises to nearly 60 percent.

More than half of this water is used either for generating electricity, or for cooling power stations, and is returned to its source virtually unchanged. Other major industrial uses of water – including chemical and petroleum plants, metal industries, the wood, pulp and paper industries, food processing and machinery manufacture – are much heavier polluters.

Industrial water use has increased only slowly since the 1980s as a result of concerted efforts to control its use and treatment. However, as more developing countries industrialize, the global industrial use of water is expected to rise steeply over the next 25 years. There is already a pressing need to develop and implement pollution control measures. In developing countries, 70 percent of industrial waste is dumped untreated into rivers, many of which have become seriously polluted. Indiscriminate industrial discharges have also contaminated soils and underground aquifers.

The amount of water used for different industrial products varies widely. Heavy water users include the paper and metal industries, and the electronics industry, which requires high-quality, treated water. The financial value generated by the use of water in industrial processes is generally greater than that from agriculture, but it also varies widely between countries.

Changing patterns of rainfall as a result of climate change may affect industrial production. In areas dependent on hydro-electricity, reduced rainfall will mean reduced volumes of surface water and therefore less power. And "green" fuels such as bio-diesel require much more water than oil-based fuels.

INCREASING INDUSTRIAL USE
Annual water withdrawals
1950–2000, 2025 projected
cubic kilometres

1,170

776

713

204

1950   1980   2000   2025

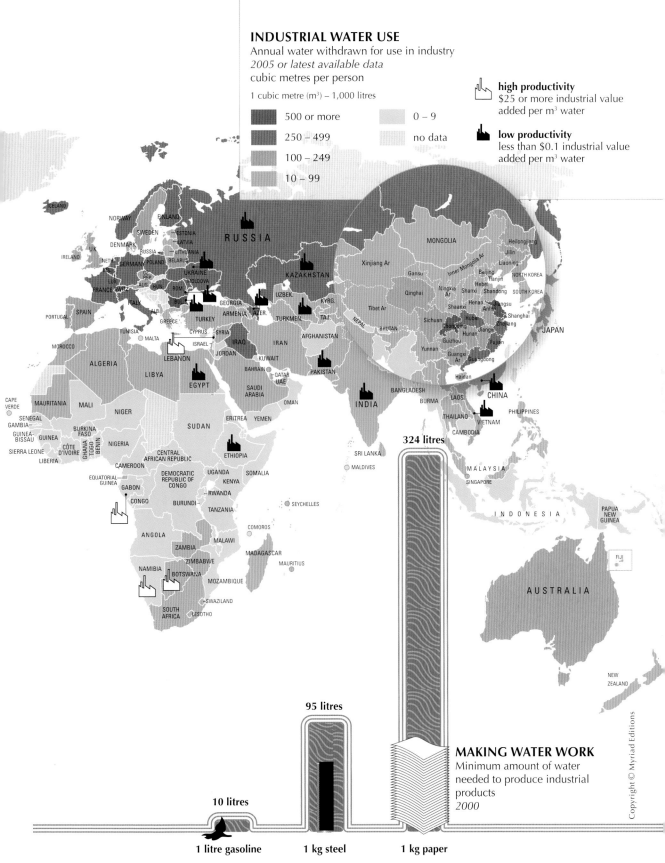

INDUSTRIAL WATER USE

Annual water withdrawn for use in industry
2005 or latest available data
cubic metres per person

1 cubic metre (m³) – 1,000 litres

- 500 or more
- 250 – 499
- 100 – 249
- 10 – 99
- 0 – 9
- no data

high productivity
$25 or more industrial value added per m³ water

low productivity
less than $0.1 industrial value added per m³ water

324 litres

95 litres

10 litres

MAKING WATER WORK
Minimum amount of water needed to produce industrial products
2000

1 litre gasoline   1 kg steel   1 kg paper

Copyright © Myriad Editions

21 Water for Energy; 25 Water Pollutants; 26 Water Pollution; 27 Damaged Waterways; 33 Water Footprint

Hydropower provides

19%

of the world's electricity

Water plays a vital role in the generation of electricity. The force of its flow is used to power turbines, and it acts as a cooling agent in gas, coal and nuclear power stations.

Hydropower is an important motivation for the construction of large dams, either as a primary purpose or as an additional function where the dam is built for irrigation or some other reason. Five countries – Canada, USA, Brazil, China and Russia – account for more than half the world's hydropower generation.

Hydropower has been promoted as a clean, low-cost, renewable source of energy, with a lighter carbon-footprint than fossil-fuel power stations, and without the risks associated with nuclear-generated energy. However, reservoirs behind large dams do release greenhouse gases, and can also lead to high levels of evaporation and consequent water-use inefficiency. Whilst some hydroelectric projects have been economically successful, others – such as the Tucurui dam in Brazil and the Pak Mun in Thailand – have failed to generate electricity as cost-effectively as less environmentally and socially damaging alternatives. A review of six hydroelectric dams in Africa revealed that four were uneconomical.

Harnessing fast, natural river flow for power generation, or building small hydropower (SHP) schemes in remote rural areas, have many benefits that avoid the financial, social and environmental costs associated with large dams. Where water is scarce, some of the electricity generated can be used to pump water into storage tanks for drinking or irrigation. China has successfully used SHP technology to benefit over 300 million rural people. Worldwide, however, such schemes currently produce only a tenth of the electricity generated by hydropower plants.

LARGE AND SMALL HYDROPOWER

Proportion of electricity generated *2007*

large hydropower 18%

small hydropower 2%

total electricity generated: 4,300 gigawatts

CANADA

USA

MEXICO

CUBA

JAMAICA HAITI DOMINICAN REP.

GUATEMALA HONDURAS
EL SALVADOR NICARAGUA N. ANTILLES

COSTA RICA TRINIDAD & TOBAGO
PANAMA VENEZUELA

COLOMBIA

ECUADOR

PERU

BRAZIL

BOLIVIA

PARAGUAY

CHILE ARGENTINA

URUGUAY

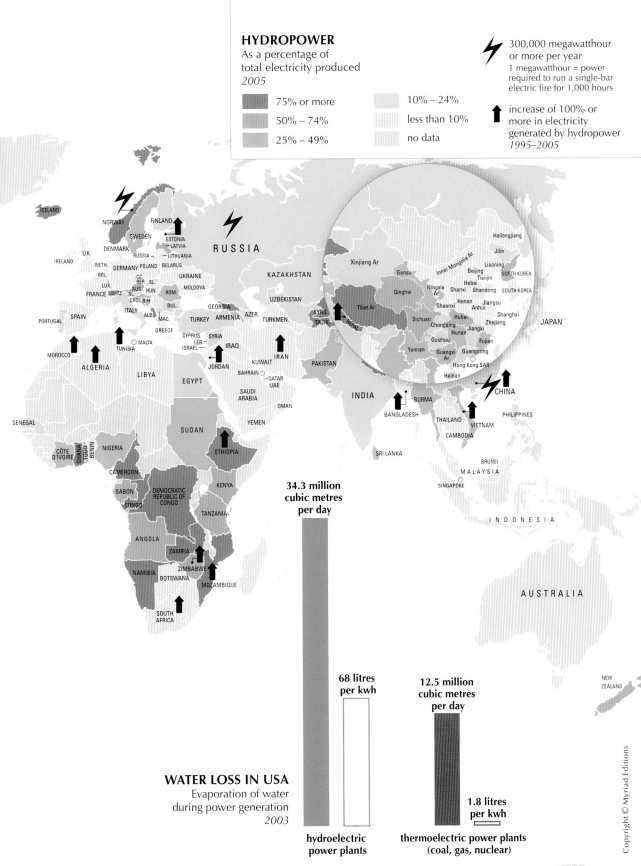

HYDROPOWER

As a percentage of
total electricity produced
2005

- 75% or more
- 50% – 74%
- 25% – 49%
- 10% – 24%
- less than 10%
- no data

300,000 megawatthour
or more per year
1 megawatthour = power
required to run a single-bar
electric fire for 1,000 hours

increase of 100% or
more in electricity
generated by hydropower
1995–2005

**34.3 million
cubic metres
per day**

**68 litres
per kwh**

**12.5 million
cubic metres
per day**

**1.8 litres
per kwh**

WATER LOSS IN USA

Evaporation of water
during power generation
2003

hydroelectric
power plants

thermoelectric power plants
(coal, gas, nuclear)

WATER FOR FISHERIES

Inland waters contributed

25%

of world fish production in 2006

Fish – in rivers, lakes and in salt-water coastal areas – make a major contribution to the global food supply, and are increasingly farmed as a cash crop.

The number of people fishing and practising aquaculture has doubled since 1970, with over 200 million dependent on it for their livelihood. The largest expansion has occurred in Asia, now the centre of the fishing industry.

Wild freshwater fish stocks have been falling dramatically, largely due to river fragmentation. Migratory fish, such as salmon, cannot pass dams, nor can other organisms critical to maintaining aquatic ecosystems. One-fifth of freshwater species are classified as extinct, endangered or vulnerable.

Fish is still the primary source of protein for millions of people in the developing world. In Bangladesh, fish provides up to 80 percent of dietary animal protein, as it does among many river- and lake-side dwellers, whose fish catches are routinely under-reported. In countries such as the Philippines and Vietnam, fish are farmed for local markets. But the increasing demand for fish and seafood in the industrialized world has led to the development, in countries such as Chile and Japan, of large commercial farms, raising high-value predatory fish such as salmon, tuna and shrimp. These require protein-rich feed, largely based on smaller fish species, many of which are being over-fished for this purpose. Aquaculture now provides at least 36 percent of the world's seafood supply.

Aquaculture has a huge impact on the environment. More than half of Asia's mangroves have been lost or degraded by the development of fish farms. The effluent and excess feed from poorly managed farms creates dead zones on lake and ocean floors. Pesticides, bleaches and antibiotics used to control diseases among intensively reared fish are washed into surrounding water, and farmed fish escape to either compete or breed with native species.

Properly guided, the explosive growth in fish farming is a hopeful trend for the world food system, but more attention needs to be paid to sustainability and standards.

INLAND WATERS
Share of fish captured
2006

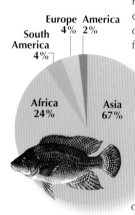

- South America 4%
- Europe 4%
- America 2%
- Africa 24%
- Asia 67%

total: 10 million tonnes

Aquaculture production increased by

80%

between 1997 and 2006

CANADA

USA

MEXICO

CUBA
JAMAICA
BELIZE
GUATEMALA HONDURAS
EL SALVADOR NICARAGUA
DOMINICAN REP. PUERTO RICO
COSTA RICA
PANAMA
VENEZUELA
GUYANA
COLOMBIA
SURINAME
ECUADOR
PERU
BRAZIL
BOLIVIA
PARAGUAY
CHILE ARGENTINA

Chile

Aquaculture production trebled in southern Chile between 1997 and 2006, and farmed salmon and trout have become a major export, bringing in over $1.6 billion a year. In 2008, however, the deadly Salmon Anemia virus, associated with industrial-scale farming, was detected at a number of sites.

 8 Altered Flows; 9 Draining Wetlands

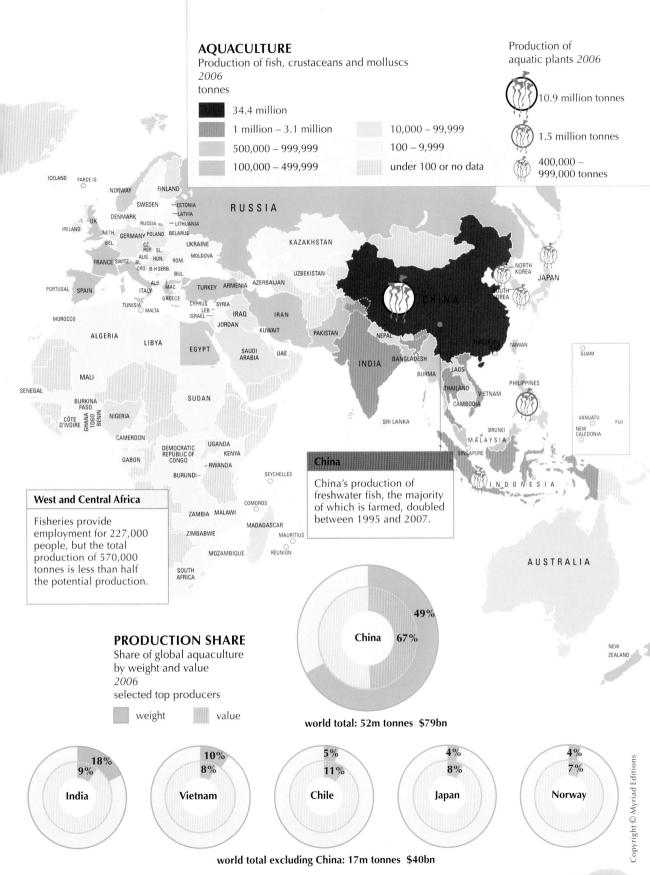

AQUACULTURE

Production of fish, crustaceans and molluscs
2006
tonnes

- 34.4 million
- 1 million – 3.1 million
- 500,000 – 999,999
- 100,000 – 499,999
- 10,000 – 99,999
- 100 – 9,999
- under 100 or no data

Production of
aquatic plants *2006*

- 10.9 million tonnes
- 1.5 million tonnes
- 400,000 – 999,000 tonnes

China

China's production of freshwater fish, the majority of which is farmed, doubled between 1995 and 2007.

West and Central Africa

Fisheries provide employment for 227,000 people, but the total production of 570,000 tonnes is less than half the potential production.

PRODUCTION SHARE

Share of global aquaculture by weight and value
2006
selected top producers

- weight
- value

China 67% 49%

world total: 52m tonnes $79bn

India 18% 9%
Vietnam 10% 8%
Chile 5% 11%
Japan 4% 8%
Norway 4% 7%

world total excluding China: 17m tonnes $40bn

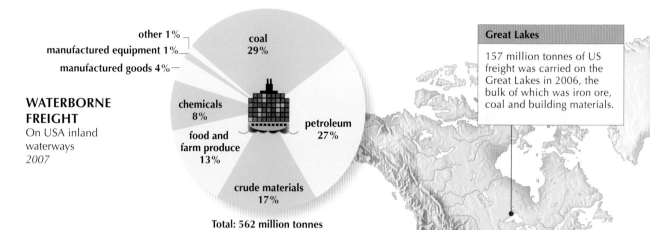

WATERBORNE FREIGHT
On USA inland waterways
2007

other 1%
manufactured equipment 1%
manufactured goods 4%
coal
29%
chemicals
8%
food and
farm produce
13%
petroleum
27%
crude materials
17%

Total: 562 million tonnes

Great Lakes

157 million tonnes of US freight was carried on the Great Lakes in 2006, the bulk of which was iron ore, coal and building materials.

Water is integral to many activities of productive and cultural significance not always measurable in economic statistics.

For centuries, waterways were more important than roads as thoroughfares and communications networks. In prehistoric times, amber was brought by river from the Baltic to southern Europe, and the Egyptians solved the problem of navigation through rapids by building a canal to bypass a cataract on the Nile as early as 2300 BC.

The Chinese completed their 1,794-km Grand Canal in 609, but the modern age of canal transport in Europe dates from 1681, when the Canal du Midi joined the Bay of Biscay to the Mediterranean. In the UK, 4,000 miles of canals were constructed before the advent of the railways which over time rendered them obsolete.

China, the USA and a few other countries still rely on navigable inland waterways for transporting freight, but elsewhere the income generated from rivers and canals comes mainly from pleasure craft. Water plays a major role in the tourist and leisure industry in several ways. Spa towns in Europe have a long history as destinations for the leisured classes. More recently, reservoirs and natural lakes have been used for sailing and other water sports.

However, pastimes such as golf, which place a heavy demand on freshwater supplies, are beginning to be called into question, especially in regions where water resources are stressed.

Golf courses in dry climates use around

8,000

litres
of water per round played

Mississippi

Total freight in 2007: 283m tonnes

downstream
177m tonnes
mainly farm produce

upstream
105m tonnes
mainly fuel

Amazon cruises

Traditional riverboats, or more overtly modern cruise ships, are part of the region's growing tourist trade and, if not regulated, are likely to have a major impact on the river's ecosystem.

European waterways

Several of Europe's largest rivers are still used to transport freight, but most of the canal networks vital to the Industrial Revolution are now almost entirely used for recreation.

Artificial beaches

In an attempt to generate more income from their rivers, several city authorities have created artificial beaches. The first was on the Seine, in Paris in 2000, but it has been imitated in London and other European cities.

China's waterways

China's 123,500 km of navigable rivers and canals are used to transport an increasing amount of freight.

2002

1.4bn tonnes

2007

2.8bn tonnes

Nile cruises

The Nile has been used for millennia to transport goods, and some of its traditional dhows are now used for tourism Most visitors, however, travel in one of the dozens of luxury cruise ships that ply the waters.

Kerala backwaters

These 900 km of interconnected lakes and canals have for centuries been used to transport goods and people, but are now being used to generate income from tourism.

Golf courses

The use of water to maintain golf courses is a hotly contested issue. Keeping 75 acres of turf healthy in a dry climate may require as much as 432 million litres of water a year.

Golf-course management practices are changing to incorporate water-saving technologies, such as rainwater harvesting, recycling of water and use of more drought-resistant grasses, but these are likely to have little overall impact in areas such as southwest USA and southern Europe, where water resources are severely stretched.

PRICE OF PIPED WATER

Cost in selected industrialized countries
2001
US$ per 200 litres

Germany
$0.38

France
$0.25

UK
$0.24

Italy
$0.15

Spain
$0.11

USA
$0.1

Australia
$0.1

South Africa
$0.09

The sale of water is an inevitable part of any organized delivery system, whether by haulers and carters, or by the most sophisticated pipeline and pump-house operation.

If people are to have taps, safe drinking supplies and flushing toilets, there has to be an industry, and the water it delivers has to be paid for by someone. However, what the price should be to different customers – farmer, industrialist, rich householder, poor slum-dweller – and who, if anyone, should benefit from its sale, remains a matter of dispute.

People living in the countryside traditionally collected their water from streams and rivers, or drew it from dug wells. Water was seen as a free natural resource and its common ownership was the basis of water laws stretching back into antiquity. But water has also been sold for domestic consumption since time immemorial, by water carriers, carters and vendors, who collect it, transport and sell it by the containerful. This trade still survives in those corners of the world where piped water has yet to be laid on by a public utility or private company.

So essential is an adequate supply of safe water to life and health that, following the industrial revolution in Europe and North America, the principle was accepted that, while customers paid "water rates", supplies should be heavily subsidized. But in many parts of the developing world where water sources are under severe stress, utilities operating on this principle cannot serve expanding populations efficiently – or even at all.

Environmental concern led to international recognition of a new principle in the 1990s, that "water's economic value should be taken into account in all its competing uses". The subsequent deployment of market mechanisms to control the use of water has been highly controversial. The poor often have to pay the full market price, while the better-off are subsidized.

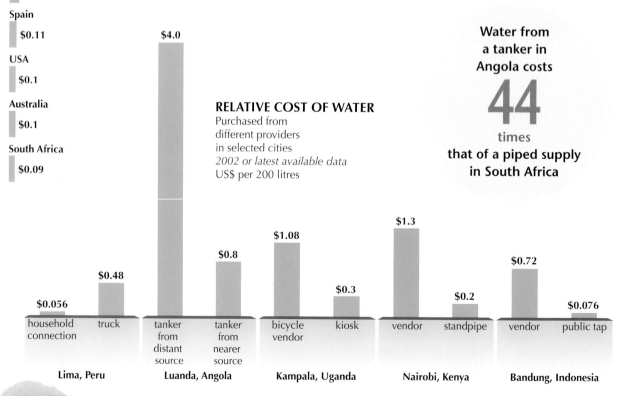

RELATIVE COST OF WATER
Purchased from different providers in selected cities
2002 or latest available data
US$ per 200 litres

Water from a tanker in Angola costs

44

times

that of a piped supply in South Africa

$4.0

$0.48

$0.056

household connection — truck

Lima, Peru

tanker from distant source

tanker from nearer source

$0.8

Luanda, Angola

bicycle vendor

kiosk

$1.08

$0.3

Kampala, Uganda

vendor

standpipe

$1.3

$0.2

Nairobi, Kenya

vendor

public tap

$0.72

$0.076

Bandung, Indonesia

BOTTLED WATER

Bottled water, once a rarefied consumer item, has become ubiquitous. Companies such as Nestlé, Danone, Coca-Cola and Pepsi have taken a natural resource, capitalized on its purportedly safe and health-preserving properties, packaged it attractively, distributed it widely, and marketed it successfully.

Even where tap water is reliably potable, people are prepared to spend up to a thousand times more on bottled water. Global sales increased at around 10 percent a year from the mid-1990s to 2005, particularly in the huge untapped Asian market. The US market alone was worth nearly $12 billion in 2007. Blind tastings have revealed, however, that many people cannot distinguish between bottled and tap water. This is not entirely surprising, since some bottled waters originate from the municipal water system. Many people continue to believe that bottled water is "cleaner", even though it may be less rigorously tested than tap water, and there have been several incidents of contamination.

The cost to the environment of bottling and packaging water is immense. Some spring waters are transported thousands of miles from their source. If the oil required to produce the plastic bottles is included, the carbon footprint of bottled water is around 600 times greater than that of tap water. Water extraction for bottling can adversely affect groundwater availability for local farmers, as shown in the landmark case, in which a Coca-Cola plant in Kerala, India was forced to close down in 2004, after a ruling that it had over-used and contaminated water supplies.

There has been a recent backlash, with public authorities cancelling contracts for bottled water. Sales may well be further affected by the economic downturn that started in 2008.

Production of plastic for US-consumed bottled water could fuel

1 million cars

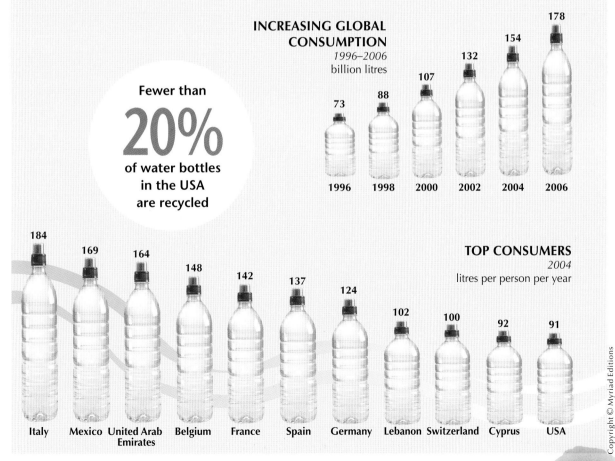

Fewer than

20%

of water bottles in the USA are recycled

INCREASING GLOBAL CONSUMPTION
1996–2006
billion litres

1996	1998	2000	2002	2004	2006
73	88	107	132	154	178

TOP CONSUMERS
2004
litres per person per year

Italy	Mexico	United Arab Emirates	Belgium	France	Spain	Germany	Lebanon	Switzerland	Cyprus	USA
184	169	164	148	142	137	124	102	100	92	91

PART 5 Damaged Water

The world's rivers and other bodies of water function as natural drainage channels and sinks. Substances deposited on the land tend to be washed into waterways by the rain, or gradually leach through the soil. Up to a point, and over time, this natural cleansing system can absorb waste matter and render it harmless. The problem is that the current volume of organic material discharged into rivers and streams is overwhelming their capacity to break it down, while other non-organic pollutants are even more toxic and cannot be absorbed.

Human excreta is one of the most serious causes of water pollution. In the developing world, around 90 percent of sewage is discharged untreated into rivers, some of which are so laden with foul matter in the dry season that aquatic life is stifled. Since rivers continue to be used for bathing, laundry and even for drinking water in many environments, the failure to treat pathogenic excreta before its discharge, or find other ways of confining it safely from human contact, represents a major public health threat.

Rivers also receive industrial wastes, and in the developing world 70 percent of these are similarly untreated. Some are organic, and eventually decompose, but not without absorbing oxygen and thereby depriving and depleting fish, plants and other aquatic creatures. Other contaminants resist environmental degradation altogether. Known as persistent organic pollutants (POPs), many of these originate from chemical fertilizers and pesticides. Their residues are washed into streams and may enter the food chain where they can build up in human tissues and have serious repercussions on health. Even more problematic are inorganic pollutants, such as heavy metals, and pharmaceutical residues that are not eliminated by conventional wastewater treatment.

Heavy industries and mines are the worst polluters of waterways. "Tailings ponds" that store toxic by-products may be accidentally spilled or overflow into the surrounding environment. Such risks are enhanced when vulnerable installations are situated near rivers or in earthquake zones, where disasters can cause serious damage to river ecosystems, killing fish and aquatic life for many miles. Pollution loads in water courses are responsible for the extinction of many freshwater species, and chemical spillages have been known to contaminate the drinking water supplies of large downstream populations.

Effective management regimes to control pollution and prevent the destruction of aquatic life and water itself are essential for environmental health. Waste discharges need to be controlled at the source wherever possible, and polluters regulated with heavy penalties.

25 WATER POLLUTANTS

HAZARDOUS WASTE

by source
2008

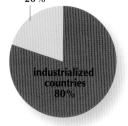

developing
countries
20%

industrialized
countries
80%

Industrial processes produce up to 500 million tonnes of waste a year. Some is discharged directly into rivers and lakes, causing surface water pollution. Other contaminants reach streams, aquifers and wetlands indirectly, collected by surface run-off or gradually leached through soils.

Globally, an important pollutant is untreated human excreta. In the developing world as a whole, around 90 percent of raw sewage – and 70 percent of industrial waste – is discharged into watercourses, creating a public health hazard for those drawing domestic water from open sources.

Organic pollutants, which include not only sewage but food-processing waste, detergents, and industrial solvents, eventually decompose in water, but in so doing absorb vital oxygen, thereby depriving fish, plants and other aquatic creatures and causing their depletion. Those that resist environmental degradation through chemical and biological processes are classified as persistent organic pollutants (POPs). Many of these originate from pesticides, and are washed into streams by rainfall. They accumulate in human and animal tissue, enter the food chain in concentrated form, and have serious repercussions on health. POPs can survive for long periods, and may be found thousands of miles from their source, carried by tidal and wind currents into pristine environments such as the Antarctic.

Inorganic pollutants, which never decompose – detritus from the construction of roads and buildings, and the heavy metals lead, cadmium and mercury, which are typical by-products of the mining industry – are also be found in the world's rivers, lakes and underground aquifers.

Where pollutants are discharged directly into waterways, there is the possibility of controlling them at the source through enforced regulation. Much harder to deal with are contaminants such as residues from cosmetics, antibiotics and other pharmaceutical products that find their way into the environment via human and animal excreta. Around 40 percent of antibiotics manufactured are fed to livestock as growth enhancers. These and many other pharmaceutical products are not eliminated by conventional wastewater treatment.

ORGANIC AND NON-ORGANIC POLLUTANTS

Pollution type	Main sources	Adverse effects
Organic matter e.g. excreta, food waste, carbon-based substances	Industrial wastewater and domestic sewage.	Decomposition leads to oxygen depletion, stressing or suffocating aquatic life.
Toxic organic compounds and micro-organic pollutants e.g. PCBs, pesticides, pharmaceuticals, solvents	Industries, motor vehicles, agriculture, gardeners, municipal waste.	Changes in oxygen levels and decomposition rate of organic matter in water, and in biodiversity.
Heavy metals e.g. cadmium, lead, zinc, copper	Industries and mining sites.	Persist in sediments and wetlands. They poison fish and pass down food chain to humans.
Pathogens and microbes e.g. cryptospiridium, salmonella, shigella	Domestic sewage, livestock.	Spread of infectious diseases and parasites.
Nutrients e.g. nitrogen and phosphorous	Run-off from agricultural lands and urban areas, industrial discharge.	Over-stimulates growth of algae, which, when they decompose, use oxygen in water, stressing or suffocating aquatic life.

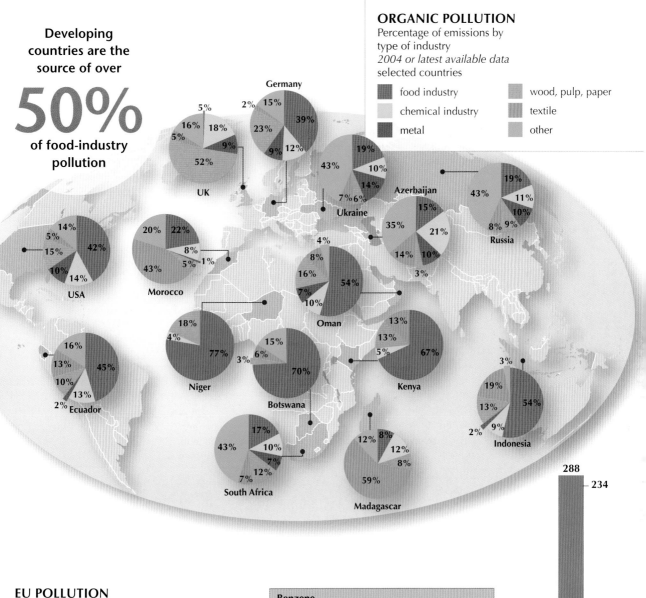

Developing countries are the source of over

50%

of food-industry pollution

ORGANIC POLLUTION
Percentage of emissions by type of industry
2004 or latest available data
selected countries

- food industry
- chemical industry
- metal
- wood, pulp, paper
- textile
- other

Germany
39% 12% 9% 23% 15% 2%

UK
52% 9% 18% 16% 5% 5%

Ukraine
43% 14% 10% 19% 7% 6%

Azerbaijan
35% 15% 21% 10% 14% 3%

Russia
43% 19% 11% 10% 9% 8%

USA
42% 14% 15% 10% 5% 14%

Morocco
43% 5% 8% 22% 20% 1%

Oman
54% 10% 7% 16% 8% 4% 3%

Kenya
67% 5% 13% 13%

Indonesia
54% 9% 2% 13% 19% 3%

Ecuador
45% 13% 10% 13% 16% 2%

Niger
77% 3% 4% 18%

Botswana
70% 6% 15%

Kenya

South Africa
43% 12% 7% 10% 17% 7%

Madagascar
59% 8% 12% 8% 12%

EU POLLUTION
Release of benzene and related chemicals by selected industries in EU25
2004
tonnes per year

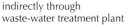 direct to water

indirectly through waste-water treatment plant

Benzene

Benzene is a chemical present in oil, gasoline and cigarette smoke. It is used in the production of plastics, resins, nylon, lubricants, detergents, drugs and pesticides. Long-term exposure affects bone marrow, and can cause anaemia and leukaemia.

combustion 3.6 — 4.9 — 1.3

hazardous waste disposal 3.3 — 7.4 — 4.1

non-hazardous waste disposal 8.2

pharmaceuticals 0.5 — 12 — 11

biocides and explosives 26

oil and gas refining 45 — 15 / 30

organic chemicals 288 — 234 / 54

The number of dead zones in the oceans increased by

30%

between 1995 and 2007

Rapid urbanization and accelerating industrialization, including mining and modern farming methods, are causing increased water pollution and corresponding environmental threats.

Human activity on an ever-more crowded planet is over-burdening the world's inbuilt self-cleansing system. The pollution of surface water in rivers, streams and lakes is one source of concern, especially where contaminants enter waterways indirectly, via run-off or seepage. But groundwater contamination is potentially even more serious if important aquifers are irreversibly damaged by the load of toxicity they bear.

Groundwater aquifers may be polluted from solid waste tips or landfill; drains from industrial sites; farmyard drainage; leaking sewers or on-site sanitation systems; or by agricultural intensification which leaves larger volumes of pesticides and fertilizer residues coating the ground. Although groundwater is less vulnerable to human impacts than surface water, cleaning it up is a lengthy and costly process, posing demanding technical problems. Methods for assessing groundwater vulnerability so as to take timely preventive action are now being given more attention.

Over-use of groundwater has had some unexpected outcomes. The lowering of the water table in West Bengal and Bangladesh has exposed seams of naturally occurring arsenic. From being a source of safe drinking water, a proportion of tube wells throughout the region – in some areas more than half – now contain water with toxic levels of arsenic which has to be filtered out. Around 13 million people are at risk in West Bengal, and 20 million in Bangladesh.

Another problem arising from the heavy pollution caused by agricultural and industrial waste is the creation of "dead zones" in waterways. The decomposition process feeds algae blooms and reduces the level of oxygen in the water to the point where no aquatic life can survive. There are 400 "dead zones" in the world's lakes and seas, covering 246,000 square kilometres altogether.

FLUORIDE

fluorosis is an endemic public health problem

Microscopic proportions of fluoride added to water improves dental health, but in excessive amounts it can cause disease and death. It occurs in the rock in certain areas, and the lowering of the water table increases its concentration. Fluorosis mainly affects bones and teeth. Its aches and pains can be relieved by increased vitamin C in the diet.

Affected piped-water can be passed through a treatment plant to eliminate fluoride. In villages where water is fetched directly from the source, it can be filtered through a household device with activated alumina granules.

Baltic Sea – coastal dead zone

Spring and summer blooms of algae are a natural occurrence in the Baltic, but nutrient-enriched river water emptying into this relatively enclosed sea worsens the problem. The nine bordering countries had a non-binding agreement to cut phosphorous emissions in half between 1989 and 2005, but Sweden is the only country to have taken the necessary steps. As well as river run-off, nitrogen emitted in the burning of fossil fuels is also a problem.

China – polluted rivers and aquifers

China's increasing population and booming economy are exacerbating its water-supply problems. With more than 75 percent of urban river water considered unfit for human consumption, people are increasingly turning to dwindling underground sources. But 90 percent of these are also believed to be contaminated with organic and inorganic pollutants, according to the country's State Environmental Protection Administration (SEPA).

Gulf of Mexico – coastal dead zone

The waters of the Mississippi drain the water from 3 million km², including some of the USA's more intensively farmed land. The nitrogen and phosphorous that is washed downriver from the fields and industries of the Midwest cause a bloom of algae at the river's mouth (shown here in red). As the algae decay, they use up oxygen, causing millions of bottom-dwelling sea creatures to die.

India – human waste

Only 20 percent of India's sewage is treated before being discharged, leading to heavily polluted rivers and lakes. In rural areas sanitation systems are mainly "on-site" with no systematic treatment of any kind. The extensive retention of excreta underground, unless carefully controlled, risks the pollution of groundwater, on which 80–90 percent of India's rural population depend for drinking water.

India – pesticide pollution

Evidence was presented in 2003 by the Centre for Science and Environment of high levels of pesticide residues in carbonated drinks produced and sold across India, including those of the major brands. The Bureau of Indian Standards subsequently revised the quality standard for such contamination, but by the end of 2008 had still not enforced it.

27 DAMAGED WATERWAYS

Industrialized society is damaging waterways. Modern lifestyles depend on mining, drilling, power-generation, chemical production, manufacturing processes and the storage and transportation of hazardous wastes.

People have always used air, land and water as "sinks" into which to dispose of waste. With industrialization, the volume and range of polluting agents has mounted, affecting the content of surface water, groundwater and rainfall itself. As countries become more industrial, their municipal, industrial and agricultural sectors cause water-quality problems that include pathogen-bearing sewage, nutrient-rich run-off that leads to a reduction in dissolved oxygen, acidification, sedimentation, and pollution by heavy metals, organic compounds and micro-pollutants,.

Industries and mining sites are the worst polluters. "Tailings ponds" are used to store waste and as depositories for toxic by-products of industrial processes; these may leach their contents into the surrounding environment, or there can be accidental spillages or embankment breaches. Acids from mining processes and power generation may also enter lakes and streams, and where they are released into the air can acidify rain. Oil and other hazardous waste spillages occasionally occur on inland waterways.

The "natech" disaster – a technological disaster triggered by a natural hazard – is likely to occur when vulnerable installations are situated close to rivers which flood, or in earthquake-prone areas. Toxic materials may be released by a strong tremor, and safety precautions dependent on spraying be rendered ineffective when water and power sources fail.

Environmental activists

Although China's government has pledged to crack down on pollution, only about 10 percent of environmental laws are enforced. Tens of thousands of environmental protests were recorded in 2007. Green Camel Bell, one of an increasing number of environmental NGOs, monitors pollution levels in the Huang He as it flows through Lanzhou and draws the attention of the authorities to offending companies.

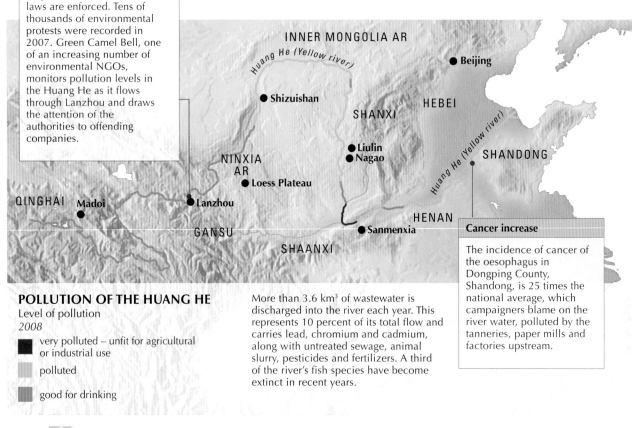

Cancer increase

The incidence of cancer of the oesophagus in Dongping County, Shandong, is 25 times the national average, which campaigners blame on the river water, polluted by the tanneries, paper mills and factories upstream.

POLLUTION OF THE HUANG HE

Level of pollution
2008

■ very polluted – unfit for agricultural or industrial use

▨ polluted

▨ good for drinking

More than 3.6 km³ of wastewater is discharged into the river each year. This represents 10 percent of its total flow and carries lead, chromium and cadmium, along with untreated sewage, animal slurry, pesticides and fertilizers. A third of the river's fish species have become extinct in recent years.

13 Water for Sanitation; 25 Water Pollutants; 26 Water Pollution

Baia Mare tailings dam, Romania

In 2000, a dam holding back mine wastewater broke, releasing up to 100,000m³ of cyanide-contamined water, as well as heavy metals into the Somes, Tisza and finally the Danube rivers and Black Sea. This interrupted the drinking-water supplies of 2.5 million people, and caused enormous damage to the river ecosystems, including the death of thousands of fish.

Songhua River, China

An explosion at a benzene factory in Jilin, in November 2005, released 100 tons of harmful chemicals into the Songhua river, a tributary of the Heilongjiang/Amur river, on the border of China and Russia. The water supplies of 4 million people in the downstream city of Harbin had to be suspended for five days.

Oil slicks on Mississippi, USA

In 2000, 1.9 million litres of crude oil was released into the river, 70 miles from the coast, when the oil tanker Westchester ran aground. In 2008, a barge leaked 1.5 million litres of heavy fuel oil into the river at New Orleans.

Earthquake, Turkey

In 1999, a powerful earthquake in northwest Turkey resulted in the release of hazardous materials, including the leakage of 6.5 million kg of toxic acrylonitrile (ACN), which contaminated air, soil and water, affecting residential areas.

Yamuna River, India

During the driest months of the year, the Yamuna river, as it flows through Delhi, consists almost entirely of wastewater, with half of the city's raw sewage being discharged into it. Despite $500 million being spent on waste-treatment stations, these are only used at two-thirds capacity. Critics point to a number of problems: the city's sewerage pipes are corroded and only "legal" sewage is sent for treatment, leaving the sewage from shanty towns to be carried untreated through open canals into the river.

DELHI'S SEWAGE TREATMENT PLANTS
Level of operating capacity of the 17 plants
2006

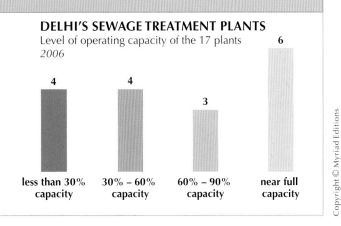

less than 30% capacity	30% – 60% capacity	60% – 90% capacity	near full capacity
4	4	3	6

Populations of freshwater species declined by

50%

between 1970 and 2000

RELATIVE SPECIES RICHNESS
Number of species compared with extent of habitat
1997

marine
0.2

terrestrial
2.7

freshwater
3.0

Even subtle changes in quality, temperature or seasonal availability of fresh water can have a devastating effect on the living organisms that inhabit it.

Freshwater environments are considered the most species-rich on earth, with complex ecologies that rely both on direct rainfall, and on water flows within and through them. They include rivers and lakes at a range of altitudes and latitudes, and many kinds of wetland.

A high proportion of plants, fish or other creatures in rivers, underground water systems, or lakes have evolved particular characteristics to suit the local ecology. Any change in that ecology threatens the extinction of these localized species.

Invasive plants and animals constitute a threat to native species, by competing for food and space, or even by killing and eating them. Canals providing navigational links between rivers spread species from one river basin to another, as do discharges of ballast water and the organisms it contains. Ecologies are also transformed by the submergence of previously farmed or forested land behind dams. River fragmentation by barrages and dams also slows flows and causes water temperatures to rise.

Non-native fish are often introduced by aquaculture, or released for recreational purposes. In over three-quarters of cases, this has led to a decline in native species. Invasive plants often crowd existing species out and reduce light and oxygen, altering the water chemistry and negatively affecting fish and other creatures. Water hyacinth, originally introduced worldwide as an ornamental species, is now found throughout the tropics, and is causing major environmental and economic problems.

Algae blooms on slow-flowing rivers or waterways filled with nutrient or pathogenic waste cause loss of species. Some algae are toxic, but even the non-toxic varieties deprive other water-dwelling organisms of oxygen when they decompose.

Biodiversity losses have been only partly detected and measured, with monitoring tending to concentrate on larger organisms. Worldwide, 20 percent of freshwater fish are vulnerable, endangered or extinct; 57 percent of freshwater otters are vulnerable or endangered, and 75 percent of freshwater molluscs in the USA are rare or imperilled. Many stocks of salmonids in western North America have been lost.

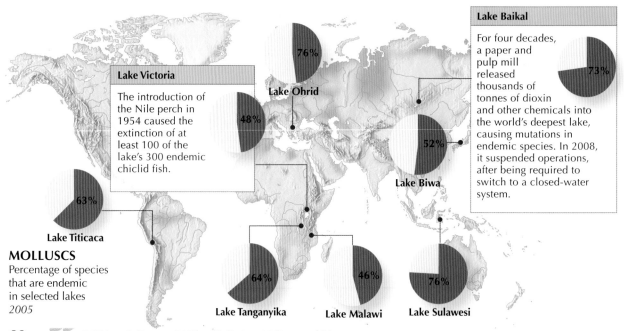

Lake Baikal

For four decades, a paper and pulp mill released thousands of tonnes of dioxin and other chemicals into the world's deepest lake, causing mutations in endemic species. In 2008, it suspended operations, after being required to switch to a closed-water system.

73%

76%

Lake Victoria

The introduction of the Nile perch in 1954 caused the extinction of at least 100 of the lake's 300 endemic chiclid fish.

Lake Ohrid
48%

52%

Lake Biwa

63%

Lake Titicaca

MOLLUSCS
Percentage of species that are endemic in selected lakes
2005

64%

Lake Tanganyika

46%

Lake Malawi

76%

Lake Sulawesi

ENDANGERED SPECIES

Species group	Status and trend	Percentage assessed by IUCN as threatened
Waterbirds and wetland-dependent birds	Birds that depend on wetland areas are more vulnerable than other species. Of the 964 wetland-dependent bird species, 203 are threatened, and 44% of populations assessed show signs of decline. Migratory birds often travel thousands of kilometres, using wetlands as vital stopping off points during which they can feed and recuperate. The disappearance or degradation of just one area can seriously affect the birds that use them. Despite international legislation requiring countries to maintain favourable conditions for waders, populations of more than half of these bird species in western Europe are declining.	21%
Wetland-dependent mammals	A range of mammals, including species of dolphin, otter, seal, shrew and hippopotamus, live or feed in fresh water. Of those assessed, over a third are threatened.	37%
Aquatic amphibians	Of 3,969 aquatic amphibians assessed, 1,045 are deemed to be threatened.	26%
Freshwater turtles	Of 200 species assessed, at least 100 are threatened. The number considered critically endangered more than doubled from 1996 to 2000. Nearly three-quarters of Asia's species are threatened, the high value placed on them for their supposed medicinal qualities contributing to their demise.	51%
Crocodiles	Of the 23 species that inhabit marshes, swamps, rivers, lagoons and estuaries, 10 are assessed as threatened.	43%
Freshwater fish	Estimates of the number of freshwater fish species vary from 10,000 to 14,000 but what is certain is that a high proportion of them are threatened with extinction. The deliberate introduction of non-native species is very often the cause.	20%
Aquatic plants	Many plants thrive in watery environments. Some live permanently underwater, or have their roots there, and others rely on seasonal flooding. There has been no systematic assessment of their status, but invasive species such as water hyacinth and Canadian pondweed clearly affect the survival of native species.	**Not assessed**

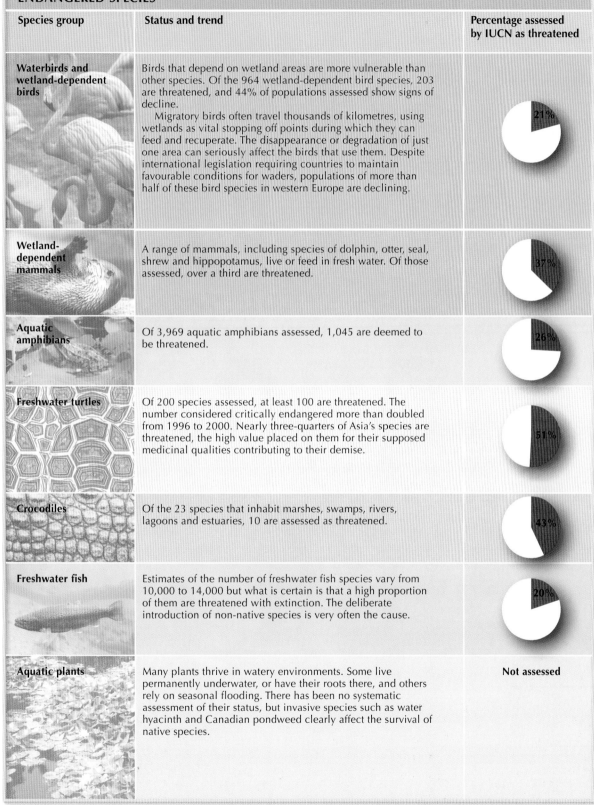

PART 6 Water for the Future

Asurprisingly large proportion of the world's population – around 1 billion people – are still without safe drinking water, and 2.5 billion are without access to sanitation. Strenuous efforts to improve water and sanitation services are underway, in keeping with the UN's Millennium Development Goals, but these efforts, alongside others to expand rain-fed and irrigated food production, take place in a world where freshwater resources are already under great stress. New tools need to be applied to ease the global water crisis.

Working together to manage and distribute fairly whatever part of this finite supply we do control is at the heart of all water-related policies. As pressures increase in the face of climate change and all the other by-products of expanding industrialized lifestyles, the priority will be for users and providers to sit down together and negotiate fair shares. This will require more water diplomacy, more activity within river basin organizations, and more attempts to treat water holistically, as a common resource that has to be efficiently and equitably shared by all its users. This in turn requires more awareness about water among politicians, professionals in different branches of water-related activity, and citizen users.

History has shown us that, faced with the need to co-operate over shared waters, people are usually up to the task. Despite fighting talk, competing parties tend to find common ground in acting to protect this vital resource. Institutional mechanisms for reconciling competing demands need to be internationally respected and strengthened if an escalation of water disputes is to be avoided in the future. In all adjudications over water, the rights of those least able to maintain their hold on the resource on which their livelihood depends need to be especially protected. At the same time, those with extravagant "water footprints" are going to have to mend their ways.

Even while ancient water-conservation systems are retooled, and new methods invented for recycling wastewater and for making saline water fresh, it is essential to understand that technological fixes are not going to save us from failure to manage our water systems. Nor is the deployment of market ideology and the profit motive any panacea for better administration of water and sanitation services or for resource conservation. Since water is essential to life itself, businesses that exploit it are not businesses like any other. Keeping this particular business clean, fair and efficient can only be done with powerful state commitment, sound regulatory regimes and real accountability.

Our Earth has been given a miraculously beneficent global water pot. Should we fail to conserve it we will have no-one to blame but ourselves.

People without decent toilets in 2008:

2.5 billion

in 2015 on current trends:

2.4 billion

No Millennium Development Goal can be met without attention to water and sanitation. Better management and access is implicit in targets for hunger and poverty reduction, public health, environmental protection and sound water governance.

In 2000, at the UN "Millennial Summit", global agreement was secured to focus international effort and resources on eight Millennium Development Goals (MDGs) that encapsulated an anti-poverty agenda. These set out measurable reductions in targets relating to poverty and development, for fulfilment by 2015. Water availability, access and affordability are essential to all the goals. Water for agriculture is required to reduce hunger, and improving the status of women requires situating water as near to their homes as possible. Water and sanitation play vital roles in hygiene and the prevention of disease. The protection of the environment, including fish stocks and aquatic life, similarly requires attention to water resources.

A specific target for drinking water was included under Goal 7 on environmental sustainability. In 2002, a rider was added to include basic sanitation. For both services, the baseline against which improvements were to be measured was coverage in 1990. The Joint Monitoring Programme of WHO and UNICEF has established indicators for measuring progress towards the drinking water and sanitation targets, and publishes reports on progress.

The 2006 data suggest that, globally, the target of halving the proportion of those without safe drinking water is likely to be reached. On sanitation, however, the world is not on track to meet the target – which was anyway far more modest than that for water, since so many fewer people in 1990 had basic toilet facilities as compared to water supplies. Another serious concern is that many reported coverage figures, especially for urban areas, are turning out to be heavily inflated. The task may be larger than anticipated.

SIGNIFICANCE OF IMPROVED WATER AND SANITATION IN MEETING MDGS

MDG	Role of water services and management
Goal 1: Eradicate extreme poverty and hunger	Improve access for subsistence farmers; raise agricultural productivity to meet demand for affordable food; make household supplies more accessible, reliable, and safer.
Goal 2: Achieve universal primary education	Provide separate in-school water and toilet blocks for staff, boys and girls, thereby reducing drop-out and disaffection by students and staff.
Goal 3: Promote gender equality and empower women	Improve women's access to water supplies, reduce time spent in water collection, and release energies for income-generating and other family support tasks.
Goal 4: Reduce child mortality	Hygiene in the home, nutritious food for infants and children, safe drinking water and correct faeces disposal have a vital influence on child illness and its outcomes.
Goal 5: Improve maternal health	Reduce risks to mother and infant by better access to safe water, especially where childbirth takes place at home; improve maternal health by better diet and hygiene.
Goal 6: Combat HIV/AIDS, malaria, and other diseases	Reduce water-related diseases by measures to control vectors, and by access to safe water, hygiene knowledge and sanitation.
Goal 7: Ensure environmental sustainability	Reverse the loss of environmental resources, and reduce by half the proportion of those without sustainable access to safe drinking water and basic sanitation.
Goal 8: Develop a global partnership for development	Practitioners, researchers and decision-makers should engage co-operatively in the integrated management of water resources.

MILLENNIUM DEVELOPMENT GOAL 7, TARGET 3

Halve, by 2015, the proportion of the population without sustainable access to safe drinking water and basic sanitation.

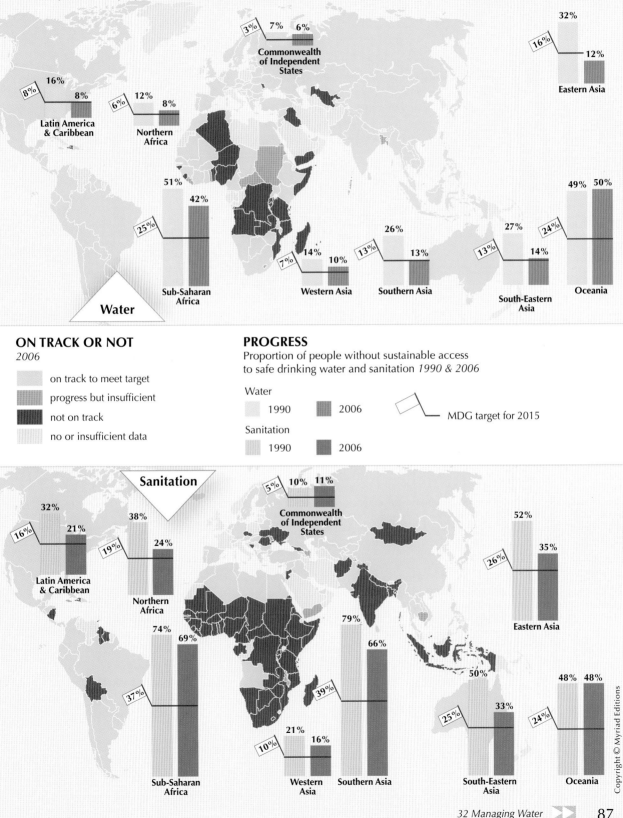

Water

ON TRACK OR NOT
2006

- on track to meet target
- progress but insufficient
- not on track
- no or insufficient data

PROGRESS

Proportion of people without sustainable access
to safe drinking water and sanitation *1990 & 2006*

Water
1990   2006   — MDG target for 2015

Sanitation
1990   2006

Sanitation

Copyright © Myriad Editions

People cannot live without water. Thus, the right to water can be seen as integral to the right to life.

Internationally, the right to water (and sanitation) is being gradually incorporated into national laws and constitutional amendments. South Africa, for example, legislated for households to receive a set volume of water free of charge. However, due to the investment required to meet a comprehensive right to water, let alone to sanitation, it has not been possible to gain universal recognition of these human rights.

More than 260 river basins are shared between countries, and equitable use of their waters requires negotiation and agreement.

The history of diplomacy over shared waters stretches back into antiquity. Unlike land, which can be demarcated, water does not stay still. Governments and legal systems have therefore accepted that the content of rivers and aquifers is a communal asset over which ownership cannot be asserted. But since political boundaries are rarely defined by river basin – in fact the opposite is true since rivers often act as boundaries – the sharing of waters between nations presents particular challenges.

Upstream fragmentation by dams, extraction of large volumes of water, and alteration in flow and water quality are among the issues requiring diplomacy, with disputes settled by treaties and binding agreements. Some international river basin organizations – such as that established for the Nile – remain in more or less permanent session, addressing new issues as they are brought to the table by the parties involved.

The development of international law concerning water has proved elusive. At the global level, there have been attempts to establish a right to water, especially since human rights became the lingua franca of development discourse in the post-Cold War era. This human right – and that to sanitation – is given some recognition through mentions in certain Conventions (such as the Convention on the Rights of the Child). Its interpretation and practical exercise in national law is under review by a Special Rapporteur to the UN Human Rights Council.

In 1997, the UN passed a Convention on the Law of the Non-Navigational Uses of International Watercourses. This was an attempt to establish overarching principles to be followed in the negotiation of watercourse treaties and the legal settlement of transboundary water disputes. This Convention has not entered international law because insufficient countries have ratified it.

13
river basins are shared by 5 or more countries

CANADA

USA

MEXICO

DOMINICAN REP.

GUATEMALA   HONDURAS
EL SALVADOR   NICARAGUA

COSTA RICA
PANAMA   VENEZUELA
COLOMBIA

○ TRINIDAD & TOBAGO

GUYANA
SURINAME

ECUADOR

PERU

BRAZIL

BOLIVIA

PARAGUAY

CHILE   ARGENTINA

URUGUAY

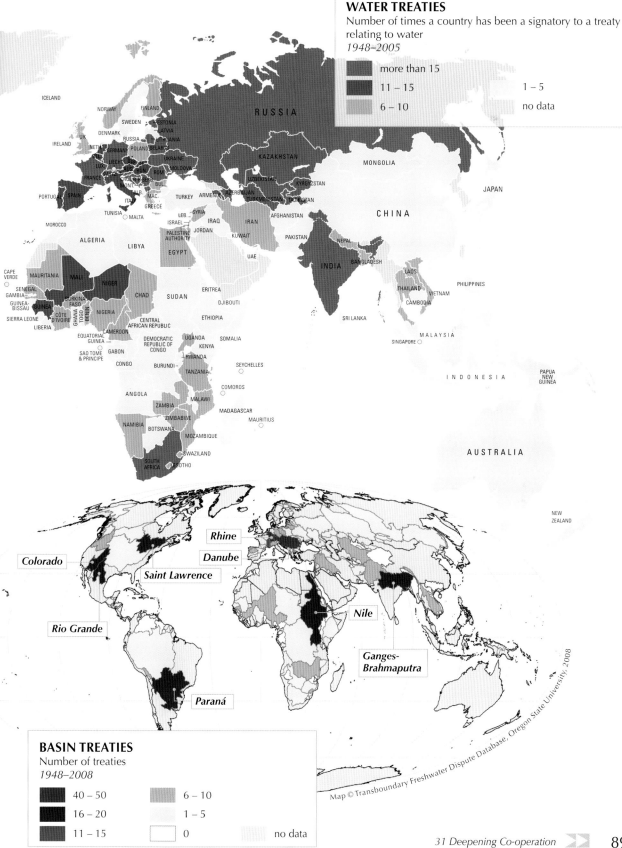

WATER TREATIES
Number of times a country has been a signatory to a treaty relating to water
1948–2005

- more than 15
- 11 – 15
- 6 – 10
- 1 – 5
- no data

Rhine

Danube

Colorado

Saint Lawrence

Rio Grande

Nile

Ganges-Brahmaputra

Paraná

Map © Transboundary Freshwater Dispute Database, Oregon State University, 2008

BASIN TREATIES
Number of treaties
1948–2008

- 40 – 50
- 16 – 20
- 11 – 15
- 6 – 10
- 1 – 5
- 0
- no data

WATER ISSUES
Co-operation and
conflict
by type of issue
1948–2008

- quantity
- joint management
- infrastructure
- quality
- hydropower
- flood control
- technical co-operation
- other

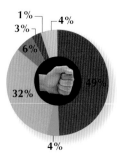

5%
4%
5%
30%
11%
8%
17%
20%

**Total co-operative events:
1,743**

1%
4%
3%
6%
49%
32%
4%

**Total conflictual events:
816**

Despite the potential for hostility over shared waters, particularly in retaliation against upstream behaviour, international co-operation is more often the rule than the exception.

The spectre of growing competition between states over water has generated fighting talk, but there is a more optimistic scenario. Managing shared waters can foster peaceful relations where they are otherwise politically strained. The Indus Waters Treaty, dividing the waters of the Indus basin between India and Pakistan, was agreed in 1960, and despite the states going to war on two occasions since then, no threat has ever been made to suspend water releases.

Managing shared waters successfully within transnational river basin organizations has tended to benefit all parties by protecting water quality, controlling floods, instituting joint monitoring and early warning systems, and agreeing wastewater emissions. But more systematic co-operation is needed.

Africa has the most politically dispersed rivers and lakes: 90 percent of all surface water in the continent is in transboundary basins, and these are inhabited by 75 percent of its people. The Nile basin is the hardest to adjudicate, with two out of 10 basin members – Egypt and Sudan – heavily dependent on a river whose final 2,700-kilometre journey is through entirely rainless desert.

River basins in the Middle East, including the Jordan, the Tigris and the Euphrates, urgently need enhanced co-operation since upstream extractions for irrigation and urban supplies are causing deep antagonism.

USA

Colorado

MEXICO

The Colorado

The Colorado river is an essential water source for western US states, and helped – contentiously – develop California's agricultural wealth. In 1922, the Colorado River Compact was signed, distributing water between upper basin states (Colorado, Wyoming, Utah) and Arizona, California and Nevada lower down. But dams, extractions and drought have so reduced flows that in some years it cannot reach the sea. The river's lowest reaches are in Mexico and the delta is highly degraded. The USA accepts responsibility for quality, and desalinates the flow close to the border, but water volumes remain uncertain.

The Danube

The Soviet Union dominated the basin from 1945 until its collapse in the early 1990s, at which point the Danube became the most internationalized basin in the world. It was also severely polluted. In 1994, the Danube Convention was signed and an international commission established. This now has 14 country members, plus the EU. Over $3 billion has been used to bring about a significant ecological recovery in the basin, and a doubling of aquatic species from 1980 levels.

Tigris–Euphrates basin

Iraq and Syria both depend on the waters of this basin for two-thirds of their water, and current development plans by Turkey for the creation of 1.7 million hectares of irrigated land could reduce flows in Syria by one-third.

The Mekong

The Mekong rises on the Tibetan plateau, and flows through six countries. More than 60 million people living in the Lower Mekong basin are dependent on its waters for food, livelihoods, transport and daily living. A Mekong River Commission was formed in 1995 between Cambodia, Laos, Thailand and Vietnam. China and Burma are "Dialogue Partners", but China is currently constructing a series of eight dams on the upper reaches without due consultation with the Commission and threatening serious impacts downstream.

The Ganges and Brahmaputra

The Ganges and Brahmaputra join the Meghna in the vast delta area of Bangladesh. The construction, by India, of a barrage at Farakka on the Ganges above Kolkata altered the flood pattern in the delta and had serious effects on food crops, water and livelihoods. After lengthy negotiations, an agreement was signed between India and Bangladesh in 1977, but lapsed in 1982. Eventually, a Ganges River Treaty was concluded in 1996 with a new formula for Farakka releases. But flows in the Ganges are lower than expected, and Bangladesh continues to protest.

The Nile

The Nile Waters Agreement (1929, re-affirmed 1954) involved Egypt and Sudan only, but decolonization in the 1960s led to the basin being shared by 10 countries. In 1999, a Nile Basin Initiative was launched with World Bank help, involving ministers from these countries, but they have not yet reached a new agreement. Downstream Egypt and Sudan are unwilling to lose their special status, and upstream countries will not accept it.

Lake Chad

Failure of the states bordering the lake to engage in joint management has contributed to an environmental disaster, leaving it one-tenth the size it was 40 years ago.

Copyright © Myriad Editions

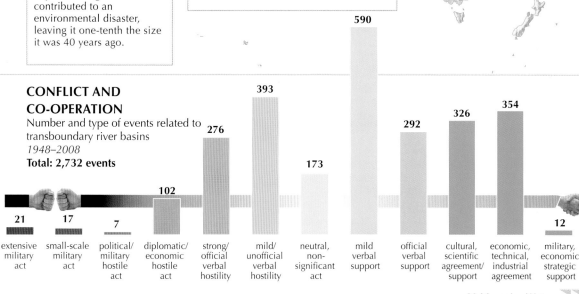

CONFLICT AND CO-OPERATION

Number and type of events related to transboundary river basins
1948–2008
Total: 2,732 events

extensive military act	small-scale military act	political/ military hostile act	diplomatic/ economic hostile act	strong/ official verbal hostility	mild/ unofficial verbal hostility	neutral, non-significant act	mild verbal support	official verbal support	cultural, scientific agreement/ support	economic, technical, industrial agreement	military, economic strategic support	international water treaty
21	17	7	102	276	393	173	590	292	326	354	12	169

The real water crisis facing the world is one of water management. Pressure on a finite resource requires efficient and equitable allocation between the rising demands of different types of user and usage.

Until quite recently, fresh water was seen as inexhaustible, and its management in different spheres of human activity – drinking water supplies, public health, agriculture, mining, industry – fragmented between government departments. But the need to treat the resource holistically – to protect its quality and also ensure that it is used economically and distributed equitably between different functions and users – led to a radical new approach: Integrated Water Resources Management (IWRM).

IWRM, first advocated in the late 1990s, emphasizes the river basin as the logical geographical unit for strategic planning – making co-operation between basin partners even more critical. The approach also encompasses other components of sustainable development: environmental protection; food security, especially for the poor; appropriate choices regarding water use in economic productivity; good governance, including decentralization of decision-making; reform of water-managing institutions; effective regulation; cost recovery and equitable pricing.

The internationally driven effort to bring all sectors concerned with water together and to enable users and providers to engage in a dialogue has proved extremely taxing. Integrated usage cuts across vested interests, requiring an approach that recognizes that water is a limited resource, and that all withdrawals and pollution affect the well-being of others. Progress towards the adoption of IWRM will take time, given different political, administrative, legal and jurisdictional systems and varying physical, social and economic conditions. The potential for corruption complicates matters, and can add substantially to the cost of implementing improvements.

Integrated management and use is often more practicable at community rather than national or international level. Water issues are felt in common locally, where everyone has a stake in conservation and efficiency, and it is striking that women are often active in local water governance. However, many communities are at the mercy of powerful interests controlling the resource, and have difficulty asserting their claims.

Corruption adds

30%

to the price of connecting to a water network in developing countries

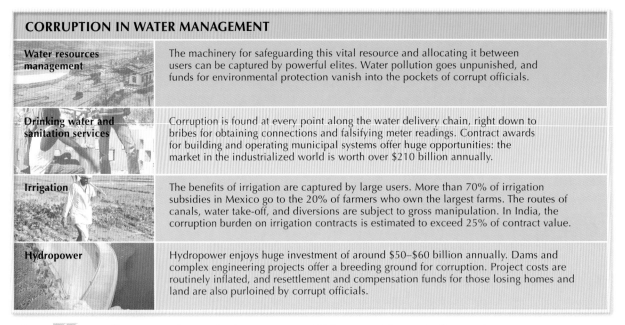

CORRUPTION IN WATER MANAGEMENT

Water resources management	The machinery for safeguarding this vital resource and allocating it between users can be captured by powerful elites. Water pollution goes unpunished, and funds for environmental protection vanish into the pockets of corrupt officials.
Drinking water and sanitation services	Corruption is found at every point along the water delivery chain, right down to bribes for obtaining connections and falsifying meter readings. Contract awards for building and operating municipal systems offer huge opportunities: the market in the industrialized world is worth over $210 billion annually.
Irrigation	The benefits of irrigation are captured by large users. More than 70% of irrigation subsidies in Mexico go to the 20% of farmers who own the largest farms. The routes of canals, water take-off, and diversions are subject to gross manipulation. In India, the corruption burden on irrigation contracts is estimated to exceed 25% of contract value.
Hydropower	Hydropower enjoys huge investment of around $50–$60 billion annually. Dams and complex engineering projects offer a breeding ground for corruption. Project costs are routinely inflated, and resettlement and compensation funds for those losing homes and land are also purloined by corrupt officials.

INTEGRATED WATER RESOURCES MANAGEMENT

Response to selective survey
by Global Water Partnership of progress
2005

- plan in place
- plan in preparation
- only initial steps taken
- not surveyed

Ferghana Valley

The soil of this once fertile valley on the borders of Uzbekistan, Kyrgyzstan and Tajikistan has become badly salinized and can no longer feed its 10 million inhabitants. Disputes have arisen over access to water resources. Water-management partnerships are now being developed across the valley, and local committees with a strong participation by women are operating more efficient water systems. Agricultural yields are increasing, and water productivity is up by 30% in some areas.

Uganda

Deterioration in quality and quantity of water has been addressed by decentralizing water management to the level of catchment areas, and the regulation of water use through a permits system.

Women and water management

Women's role in managing drinking water and sanitation services, including carrying out equipment maintenance and minor repairs, can be important.

Village women normally have to collect household water and deal with sanitation and waste disposal, so they have a greater vested interest in service performance than men, and frequently take a lead in local water and sanitation committees. The introduction of sound water management practices in low-income areas should therefore include women's participation.

Local water management

The village of Thunthi Kankasiya in Gujarat, India is an example of successful IWRM at local level. During the 1990s it developed its own micro-watershed area, damming a river, and digging 23 wells, thereby providing a year-round water supply and enabling the irrigation of 135 hectares of land. The villagers were able to grow up to three crops a year, and quadrupled the production per hectare. This raised average annual household incomes from 8,950 to 35,620 rupees, and significantly reduced the need for people to leave the area for months at a time for work.

hectares under cultivation

85

before 1991

153

2001

CONTRASTING WATER PROFILES

Proportion of embedded water by source
2001 or latest available data

- ☐ national source
- ☐ external source
- ▨ agriculture
- ▨ industry
- ▨ domestic

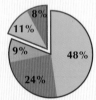

8%
11%
9%
48%
24%

USA

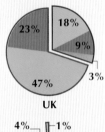

23%
18%
9%
47%
3%

UK

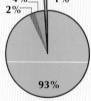

4%
1%
2%
93%

India

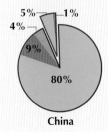

5%
1%
4%
9%
80%

China

Industrialized lifestyles, benefiting from imported as well as locally produced goods, involve the consumption of "virtual" water embedded in food and manufactured items.

Everything we eat and use – cars, computers and industrial machinery – contains water, or has required water for its manufacture. Sophisticated diets are more water-consumptive than subsistence ones, with vastly more water needed to produce meat than grains. International trade thus contains embedded water.

The concept of the "water footprint" measures the water embedded in the food, products and services that contribute to consumption patterns. It takes into account water embedded in imported as well as locally produced goods, with water embedded in exports being deducted and added to the total for the importer. A map showing water footprints thus looks very different from one that shows direct water use alone and ignores the invisible "virtual" water trade.

Several factors contribute to a large water footprint, of which industrialized lifestyle is the most obvious. However, food production in a hot climate in which moisture rapidly evaporates from the soil requires more water than in a temperate zone, accounting for the large footprints in some tropical countries. Inefficient water use, including leakage and pollution, also enlarges the overall footprint.

The concepts of virtual water and water footprints have provoked a debate about whether global water supplies could be conserved by discouraging the production and export of heavily water-embedded crops and goods from water-short areas, and encouraging water-dependent economic activity to be situated in water-rich areas. The problem is that the development of most production, export and trade is driven by economic and political considerations that are currently looming much larger than the issue of water usage.

CANADA

USA

MEXICO

CUBA

DOMINICAN REP

JAMAICA   HAITI
BELIZE
GUATEMALA   HONDURAS
EL SALVADOR   NICARAGUA

COSTA RICA

BARBADOS
TRINIDAD & TOBAGO

PANAMA   VENEZUELA   GUYANA

COLOMBIA   SURINAME

ECUADOR

PERU

BRAZIL

BOLIVIA

PARAGUAY

CHILE   ARGENTINA

16%

of water used
is for
the production
of food or
goods for
export

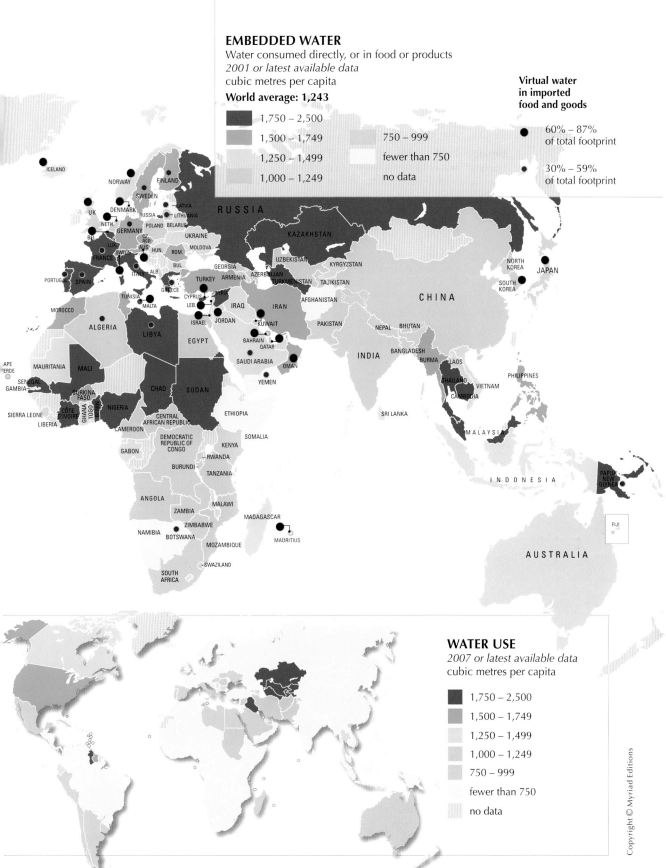

EMBEDDED WATER

Water consumed directly, or in food or products
2001 or latest available data
cubic metres per capita

World average: 1,243

- 1,750 – 2,500
- 1,500 – 1,749
- 1,250 – 1,499
- 1,000 – 1,249
- 750 – 999
- fewer than 750
- no data

Virtual water in imported food and goods

- 60% – 87% of total footprint
- 30% – 59% of total footprint

WATER USE

2007 or latest available data
cubic metres per capita

- 1,750 – 2,500
- 1,500 – 1,749
- 1,250 – 1,499
- 1,000 – 1,249
- 750 – 999
- fewer than 750
- no data

95

34 WATER AT A PRICE

Water, despite its status as a vital resource to which everybody has a right, is now seen as a commodity for which a realistic price should be paid. But what should that price be, and who should pay it?

There has been a mounting realization that the world's freshwater resources are being squandered, and need to be conserved. In the 1990s, the idea of doing this by means of pricing became conjoined with market ideology, which supported the privatization of public institutions. Politicians and financiers therefore promoted the privatization of public utilities on the assumption that operation according to market principles would simultaneously conserve water, improve efficiency and increase service spread. Subsidies would end, and all consumers – domestic, agricultural and industrial – would be charged for water at the price it cost to capture, treat and deliver it. Thus, water became a commodity from which profits could and should be made.

The UK led the way, with full-scale privatization of its water-related assets. Other countries followed, and privatization of utilities in developing countries became a condition of World Bank–IMF structural adjustment programmes and loans. Many utilities in developing countries did need reforming: they were inefficient, corrupt, and failed to provide services to poorer citizens. But privatization and partnership from the burgeoning international water corporations was no panacea. Water services, it transpires, could not easily be treated as other businesses.

The primary reason is political. The price hikes needed to reach full cost-recovery – whether from farmers drawing water for irrigation or from domestic consumers – are not viable. Protests erupt, not only from poorer populations, but from local councils and administrations. The complications of building sound and equitable water-management systems cannot be solved simplistically by market reforms, especially in environments with many vested, corruption-prone and unaccountable interests. Some international water corporations have now retreated, often in disarray, defeated by the many complexities and insufficient margins for profitability.

Privatization or no privatization, the challenge of spreading affordable services to low-income consumers persists, as does the need to address water conservation by all available means, including controlling subsidies and sensible pricing.

AS SUPPLIES DIMINISH, COSTS INCREASE

As water tables drop and sources close to urban centres dry up, increasing power is needed to pump water from deeper levels or more distant sources.

Mexico City, having over-drained the Mexico Valley aquifer, is now forced to pump water a distance of 180 kilometres and up 1,000 metres from the Cutzamala River. In so doing, it uses enormous amounts of electricity, adding substantially to the cost of the water.

The underground aquifers that supply Jakarta have been so depleted that seawater has seeped 15 kilometres inland, making the water saline. Investments of at least $1 billion are needed in pipelines to bring water from elsewhere.

The more powerful pumps needed to raise deeper water increase the cost, or make the task impossible. Small-scale farmers in developing countries may find their cheap pumps no longer up to the job. US farmers and water companies are more likely to make the necessary investment, but may still find that the additional cost renders crops uneconomic.

Power plants dependent on water to cool them may need to make expensive adjustments to their water intakes to cope with dropping water levels. Diminished water flow will reduce the amount of electricity generated by hydropower plants, making it more expensive.

$3 – $70

COST OF POWER FOR WATER EXTRACTION
USA *2003*
US$ $ per hectare

$4 – $33

pumping groundwater

lifting or pressurizing surface water

France

The two largest water corporations in the world are Suez and Véolia. As the Compagnie Lyonnaise des Eaux and Compagnie Général des Eaux they controlled water services in France for over 100 years. They expanded internationally in the market-dominated 1990s, but have since fallen into disrepute. In 1999, Suez executives were given prison sentences for corrupt practices over the privatization of services in the city of Grenoble. The contract with Suez was cancelled and returned to the municipality, and the price of water quickly became one of the lowest in France. The Mayor of Paris has announced that the city's contracts with Suez and Véolia will end after 2009, and that the management of Parisian water will become publicly controlled.

SUEZ WATER
Number of people supplied
2004–07

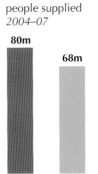

80m
68m

2004–05   2006–07

Turkey

The Turkish government is pushing through one of the most sweeping water privatization programmes in the world. Not only will all public services be privatized, but lakes and rivers will be sold off to private companies for up to 49 years. This will require a constitutional amendment: Article 43 currently limits private control of water sources and prioritizes the public interest. The privatization of irrigation water is being strenuously resisted by small-scale farmers, and popular movements in towns – often led by women – are protesting against extortionate company practices. Suez, the French water giant, pulled out of a contract in the city of Antalya after the municipality rejected their demand for another price rise, on top of an initial rise of 130%.

Tanzania

Privatization of water services was imposed on the Tanzanian government as a condition of IMF and World Bank debt relief and concessionary loans, and UK aid was used to support the privatization process. A subsidiary of the UK company Biwater took on the contract in Dar es Salaam in 2003. The Tanzanian government cancelled the contract after less than two years, citing failure to meet the contract targets. A Biwater company then took the government to the International Centre for the Settlement of Investment Disputes court and sought $20 million damages for breach of contract. The case failed.

Bolivia

The most famous case of protest against water privatization took place in Cochabamba, Bolivia, in 1999–2000. A local company, Aguas del Tunari, owned by the US

company Bechtel, was given a monopoly to collect water charges. The company took over water systems constructed and run by householders and small independent co-operatives without compensation. Prices shot up and citizens were forced to pay for collecting their own rain. Months of simmering protest led to the occupation of the city square by 80,000 people. The government sent in troops. Street battles ensued and Aguas del Tunari fled the country.

Guyana

In early 2007, the Guyanese authorities cancelled a five-year water management contract with Severn Trent Water International (STWI). The UK aid budget had been used to pay company fees for the contract and for privatization consultants, thanks to policy commitment to utility privatization and favour for home-based companies. The Guyanese audited STWI's performance, and found that only a tiny proportion of those supposed to receive potable water by 2005 had done so. Altogether, the company failed to meet five out of seven objectives in the contract. Subsequently, STWI pulled out of a bid from a similar contract in Nepal.

WATER SUPPLY FOR AMERINDIAN SETTLEMENTS

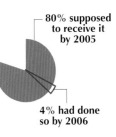

80% supposed to receive it by 2005

4% had done so by 2006

35 TECHNOLOGICAL FIXES

Innovative check-dams

Check-dams are built by farmers throughout the sub-continent to conserve rainwater after the monsoon and use it for irrigation. They are typically built with stones and mud; the idea of using plastic sheets over bare sand is an innovation. Community labour is often used, and everyone benefits from the raised water table.

Technological innovation and adaptations have a role in meeting the mounting threats to freshwater supplies that should not be overlooked – nor overstated.

Pressure on water resources has led in some settings to extraordinary responses. In India, trains full of water are routinely despatched to drought areas and Spain has similarly been forced to import water in tankers. Some small islands do the same. In such circumstances, the idea of futuristic technological rescue from water shortage – towing ice-bergs to desert areas, seeding rain, filtration and re-use at household level – has strong appeal. Meanwhile, hydro-geologists drill for deeper aquifers, engineers plan ever-larger pipelines and canals, and inventors offer tug-drawn plastic pods for shifting water by sea.

Some of the most viable technological approaches in water-short rain-fed agricultural zones have been practised for centuries, but were pushed out by industrialization and its "big irrigation project" mentality. Rainwater harvesting – using small check-dams in riverbeds, underground cisterns and roof-top collection tanks – has enjoyed a wide-scale renaissance in India after forceful campaigning by water activists. Sound local water resource management and appropriate technology here go hand in hand.

The conversion of seawater to fresh water has recently moved beyond alchemy to reality. Technologies have improved significantly in terms of performance and energy efficiency and, despite its expense, desalination is being increasingly adopted in seriously water-short settings. Estimates of the worldwide volume of desalinized water vary, but even the highest estimate of 55.6 million cubic metres a day still represents only 0.5 percent of global water use.

The main disadvantage of desalination is its heavy power consumption. Reverse osmosis, the most power-efficient method, uses between 4 and 25 kWh per cubic metre of fresh water, and thermal distillation considerably more. Although a plant using wind-generated electricity has been built near Perth, Australia, desalination plants tend to use fossil fuels, and their adoption therefore runs counter to the need to reduce greenhouse gas emissions.

USA

Every state now has at least one desalination plant.

There are over

13,000

desalination plants in the world

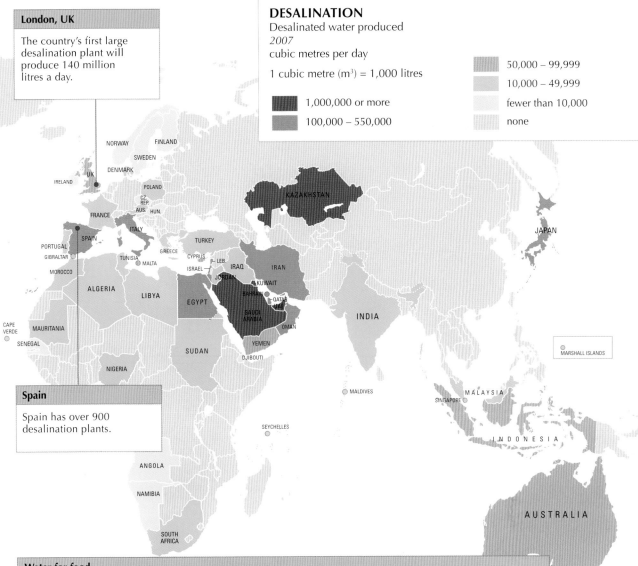

DESALINATION

Desalinated water produced
2007
cubic metres per day

1 cubic metre (m³) = 1,000 litres

50,000 – 99,999	
10,000 – 49,999	
1,000,000 or more	
fewer than 10,000	
100,000 – 550,000	
none	

Spain

Spain has over 900 desalination plants.

Water for food

The more efficient use of water to grow food is the focus of many technological developments. While research is directed at creating drought-resistant plants, farmers are being encouraged to use irrigation methods that deliver water directly to the base of plants, rather than spraying it into the air, and to enhance the soil quality so that it holds more water.

In hot, dry countries, crops need much more water than in temperate ones. Instead of desalinating seawater, an innovatory solution is to use it to cool and humidify the air inside a specially equipped greenhouse, thereby reducing the amount of fresh water needed for irrigation by up to 8 litres of water a day for every square metre of crops, and relieving pressure on local water resources.

PART 7 Water data

In recent years, extraordinary efforts have been made by international agencies to improve the quality of information available on all sorts of subjects, including water. Decision-makers increasingly insist that policies and programmes must be "evidence-based"; yet one of the characteristics of a "developing" country, or an underdeveloped area within a better-off country, is that evidence is lacking. So at the same time as trying to make sound decisions about where and how to spend resources, on water as well as in all sorts of other human development contexts, it is also necessary to improve the means of collecting and analyzing the data on which such decisions have to be made.

With the best will in the world, water data are still far from perfect. For one thing, the range of subject areas – from meteorology to health, from industrial pollution to irrigation – is so vast. Often, attempts to collect better information – as has been done by UNESCO's World Water Assessment Programme for presentation in the UN's World Water Reports – reveal that we know far less about a subject than we thought. Accurate statistics about the rate of loss of freshwater fish species, for example, or the number of people annually affected by water-related disease, turn out to be elusive. The Joint Monitoring Programme of WHO and UNICEF, assessing progress towards the fulfilment of the water and sanitation Millennium Development Goals, has discovered many anomalies in the way countries collect and report service data.

The purposes of information-gathering and presentation differ, and information may carry a political charge. The discovery of arsenic in groundwater in Bangladesh caused a national furore: who covered this information up or failed to address it in good time? Similarly, estimates of water quantities flowing in major rivers and the amount of take-off by upstream users have huge potential for dispute. Tallies of how many people are losing out on access to standpipes or toilets may deliberately minimize the number needing improvements. Populations squatting illegally on land designated as "agricultural" may be left out of account altogether, even though they have been there for decades. In all cases, averages – of water consumption, water use in agriculture, pollutant discharge – on a per capita basis by country distort the real state of affairs for almost everyone.

Any presentation of data in the format of an atlas is additionally circumscribed by the requirement to express information in a way to which it may not easily lend itself. Water primarily occurs in river basins, not in countries. Water's physical presence defines human settlement, economic activity, and patterns of life. Yet all the data connected to the availability of the resources, and its use and misuse, reflects an entirely different and often unconnected arrangement of human society in nation states. Water is truly one of the most pervasive, yet elusive, substances on the planet, and one of the hardest to understand on a country-by-country basis, both statistically and analytically.

Needs and Resources

	1 Total population millions		2 Urban population as % of total		3 Improved water source % of population with access 2004–06	4 Improved sanitation % of population with access 2004–06
	2007	2025 projected	2010	2025 projected		
Afghanistan	27.15	46.93	25%	33%	22%	30%
Albania	3.19	3.49	49%	60%	97%	97%
Algeria	33.86	42.88	67%	74%	85%	94%
Angola	17.02	27.32	57%	66%	51%	50%
Antigua and Barbuda	0.09	0.1	42%	52%	91%	95%
Argentina	39.53	46.12	91%	93%	96%	91%
Armenia	3	2.91	64%	67%	98%	91%
Australia	20.74	24.39	89%	91%	100%	100%
Austria	8.36	8.62	67%	71%	100%	100%
Azerbaijan	8.47	9.51	52%	57%	78%	80%
Bahamas	0.33	0.4	92%	93%	97%	100%
Bahrain	0.75	0.97	98%	99%	–	–
Bangladesh	158.67	206.02	27%	36%	80%	36%
Barbados	0.29	0.3	56%	65%	100%	99%
Belarus	9.69	8.67	75%	80%	100%	93%
Belgium	10.46	10.74	97%	98%	–	–
Belize	0.29	0.39	49%	56%	91%	47%
Benin	9.03	14.46	42%	51%	65%	30%
Bhutan	0.66	0.82	13%	20%	81%	52%
Bolivia	9.53	12.37	67%	73%	86%	43%
Bosnia and Herzegovina	3.94	3.75	49%	59%	99%	95%
Botswana	1.88	2.27	61%	70%	96%	47%
Brazil	191.79	228.83	87%	90%	91%	77%
Brunei	0.39	0.53	76%	81%	–	–
Bulgaria	7.64	6.55	71%	76%	99%	99%
Burkina Faso	14.78	23.73	20%	29%	72%	13%
Burma	48.8	55.37	34%	45%	80%	82%
Burundi	8.51	15.04	12%	18%	71%	41%
Cambodia	14.44	19.49	23%	33%	65%	28%
Cameroon	18.55	25.14	59%	69%	70%	51%
Canada	32.88	37.91	81%	83%	100%	100%
Cape Verde	0.53	0.75	61%	70%	80%	43%
Central African Republic	4.34	5.83	39%	45%	66%	31%
Chad	10.78	17.5	28%	38%	48%	9%
Chile	16.64	19.27	89%	92%	95%	94%
China	1,328.63	1,445.78	45%	57%	88%	65%
Colombia	46.16	55.56	74%	79%	93%	78%
Comoros	0.84	1.22	40%	51%	85%	35%
Congo	3.77	5.36	62%	69%	71%	20%
Congo, Dem. Rep.	62.64	107.48	35%	46%	46%	31%
Cook Islands	0.01	0.01	–	–	–	100%
Costa Rica	4.47	5.55	64%	72%	98%	96%
Côte d'Ivoire	19.26	26.23	47%	56%	81%	24%
Croatia	4.56	4.27	58%	64%	99%	99%
Cuba	11.27	11.23	75%	76%	91%	98%
Cyprus	0.86	1.02	70%	75%	100%	100%
Czech Republic	10.19	9.91	74%	76%	100%	99%
Denmark	5.44	5.58	86%	88%	100%	100%

5 Total renewable water resources m³ per person per year 2003–07	6 Water dependency % of renewable water originating outside country 2007	7 Annual renewable surface water million m³ 2007 or latest	8 Annual renewable ground water million m³ 2007	9 Desalinated water produced million m³ per day 2007	
2,177	15%	10,000	0	–	Afghanistan
13,323	35%	37,850	6,200	–	Albania
355	4%	10,150	1,517	17.0	Algeria
9,284	0%	145,000	58,000	0.1	Angola
642	0%	0	0	3.3	Antigua and Barbuda
21,008	66%	814,000	128,000	0.7	Argentina
3,491	14%	7,729	4,200	–	Armenia
24,411	0%	440,000	72,000	29.8	Australia
9,488	29%	77,700	6,000	2.2	Austria
3,599	73%	28,115	6,510	–	Azerbaijan
62	0%	0	0	7.4	Bahamas
160	97%	4	112	102.4	Bahrain
8,536	91%	1,189,520	21,122	–	Bangladesh
296	0%	8	74	14.6	Barbados
5,946	36%	58,000	18,000	–	Belarus
1,756	34%	18,300	900	–	Belgium
68,722	14%	0	0	–	Belize
3,128	61%	26,093	1,800	–	Benin
43,920	0%	95,000	0	–	Bhutan
67,799	51%	596,407	130,000	–	Bolivia
9,598	5%	0	0	–	Bosnia and Herzegovina
6,935	80%	10,640	1,700	–	Botswana
44,167	34%	8,233,000	1,874,000	–	Brazil
22,727	0%	8,500	100	–	Brunei
2,757	1%	20,400	6,400	–	Bulgaria
945	0%	8,000	9,500	–	Burkina Faso
20,697	16%	1,039,601	156,000	–	Burma
1,661	20%	12,536	7,470	–	Burundi
33,836	75%	471,510	17,600	–	Cambodia
17,492	4%	280,500	100,000	–	Cameroon
89,934	2%	2,892,000	370,000	2.7	Canada
592	0%	181	124	1.7	Cape Verde
35,760	2%	144,400	56,000	–	Central African Republic
4,411	65%	41,500	11,500	–	Chad
56,582	4%	922,000	140,000	4.8	Chile
2,138	1%	2,728,669	828,800	–	China
46,754	1%	2,132,000	510,000	2.7	Colombia
1,504	0%	200	1,000	–	Comoros
227,557	73%	832,000	122,000	–	Congo
22,294	30%	1,282,000	421,000	–	Congo, Dem. Rep.
–	–	0	0	–	Cook Islands
25,976	0%	75,100	37,300	–	Costa Rica
4,470	5%	78,300	37,840	–	Côte d'Ivoire
23,182	64%	95,000	11,000	–	Croatia
3,383	0%	31,640	6,480	6.9	Cuba
934	0%	560	410	33.5	Cyprus
1,287	0%	13,150	1,430	0.2	Czech Republic
1,105	0%	3,700	4,300	15.0	Denmark

Needs and Resources

	1 Total population millions		2 Urban population as % of total		3 Improved water source % of population with access 2004–06	4 Improved sanitation % of population with access 2004–06
	2007	*2025 projected*	*2010*	*2025 projected*		
Djibouti	0.83	1.11	88%	91%	92%	67%
Dominica	0.07	0.07	75%	80%	97%	84%
Dominican Republic	9.76	12.17	71%	78%	95%	79%
East Timor	1.16	2.01	29%	37%	62%	42%
Ecuador	13.34	16.07	65%	72%	95%	84%
Egypt	75.5	98.51	44%	51%	98%	66%
El Salvador	6.86	8.53	61%	67%	84%	86%
Equatorial Guinea	0.51	0.77	40%	46%	43%	51%
Eritrea	4.85	7.68	22%	31%	60%	5%
Estonia	1.34	1.25	69%	73%	100%	95%
Ethiopia	83.1	125	17%	24%	42%	11%
Fiji	0.84	0.91	53%	62%	47%	71%
Finland	5.28	5.46	62%	67%	100%	100%
France	61.65	65.77	78%	82%	100%	–
Gabon	1.33	1.7	86%	90%	87%	36%
Gambia	1.71	2.53	58%	68%	86%	52%
Georgia	4.4	3.95	53%	58%	99%	93%
Germany	82.6	80.34	76%	79%	100%	100%
Ghana	23.48	31.99	52%	62%	80%	10%
Greece	11.15	11.24	60%	65%	100%	98%
Grenada	0.11	0.11	31%	37%	95%	97%
Guatemala	13.35	19.93	50%	58%	96%	84%
Guinea	9.37	14.55	35%	45%	70%	19%
Guinea-Bissau	1.7	2.91	30%	35%	57%	33%
Guyana	0.74	0.68	29%	34%	93%	81%
Haiti	9.6	12.31	42%	53%	58%	19%
Honduras	7.11	9.68	49%	57%	84%	66%
Hungary	10.03	9.45	68%	74%	100%	100%
Iceland	0.3	0.34	93%	94%	100%	100%
India	1,169.02	1,447.50	30%	37%	89%	28%
Indonesia	231.63	271.23	54%	66%	80%	52%
Iran	71.21	88.03	70%	76%	94%	–
Iraq	28.99	43.29	67%	69%	77%	76%
Ireland	4.3	5.28	62%	68%	–	–
Israel	6.93	8.72	92%	93%	100%	–
Italy	58.88	58.08	68%	73%	–	–
Jamaica	2.71	2.91	55%	62%	93%	83%
Japan	127.97	121.61	67%	72%	100%	100%
Jordan	5.92	8.03	84%	88%	98%	85%
Kazakhstan	15.42	16.99	59%	64%	96%	97%
Kenya	37.54	57.18	22%	30%	57%	42%
Kiribati	0.1	0.12	52%	62%	65%	33%
Korea, North	23.79	25.23	63%	70%	100%	59%
Korea, South	48.22	49.02	82%	85%	92%	–
Kuwait	2.85	3.99	98%	99%	–	–
Kyrgyzstan	5.32	6.21	37%	43%	89%	93%
Laos	5.86	7.71	23%	31%	60%	48%
Latvia	2.28	2.07	68%	72%	99%	78%

5 Total renewable water resources m³ per person per year 2003–07	6 Water dependency % of renewable water originating outside country 2007	7 Annual renewable surface water million m³ 2007 or latest	8 Annual renewable ground water million m³ 2007	9 Desalinated water produced million m³ per day 2007	
378	0%	300	15	0.1	Djibouti
–	0%	0	0	–	Dominica
2,360	0%	20,995	11,695	–	Dominican Republic
–	–	0	0	–	East Timor
32,083	0%	424,400	134,000	2.2	Ecuador
787	97%	56,000	1,300	100.0	Egypt
3,667	30%	25,080	6,150	0.1	El Salvador
51,587	0%	25,000	10,000	–	Equatorial Guinea
1,431	56%	6,200	500	–	Eritrea
9,630	1%	11,808	4,000	–	Estonia
1,576	0%	120,000	20,000	–	Ethiopia
33,667	0%	0	0	–	Fiji
20,956	3%	109,800	2,200	0.1	Finland
3,367	12%	201,700	100,000	11.7	France
118,497	0%	162,000	62,000	–	Gabon
5,274	63%	8,000	500	–	Gambia
14,155	8%	62,100	17,230	–	Georgia
1,862	31%	153,300	45,700	–	Germany
2,406	43%	51,900	26,300	–	Ghana
6,677	22%	71,750	10,300	10.0	Greece
–	0%	0	0	–	Grenada
8,832	2%	102,770	33,700	–	Guatemala
24,037	0%	226,000	38,000	–	Guinea
19,546	48%	27,000	14,000	–	Guinea-Bissau
320,905	0%	241,000	103,000	–	Guyana
1,645	7%	11,868	2,157	–	Haiti
13,314	0%	86,920	39,000	0.2	Honduras
10,299	94%	104,000	6,000	0.2	Hungary
576,271	0%	166,000	24,000	–	Iceland
1,719	34%	1,858,120	418,540	0.5	India
12,739	0%	2,793,000	455,000	18.7	Indonesia
1,978	7%	106,310	49,300	200.0	Iran
2,618	53%	74,220	3,200	7.4	Iraq
12,536	6%	51,200	10,800	0.2	Ireland
248	55%	595	1,075	140.0	Israel
3,293	5%	179,300	43,000	97.3	Italy
3,547	0%	5,512	3,892	0.5	Jamaica
3,357	0%	420,000	27,000	40.0	Japan
154	23%	685	500	9.8	Jordan
7,394	31%	103,510	6,100	1328.0	Kazakhstan
896	33%	30,200	3,500	–	Kenya
–	–	0	0	–	Kiribati
3,430	13%	76,135	13,000	–	Korea, North
1,458	7%	67,100	13,300	–	Korea, South
7	100%	0	20	420.2	Kuwait
3,910	0%	18,180	13,600	–	Kyrgyzstan
56,305	43%	333,550	37,900	–	Laos
15,366	53%	35,249	2,200	–	Latvia

Needs and Resources

	1 Total population millions		2 Urban population as % of total		3 Improved water source % of population with access 2004–06	4 Improved sanitation % of population with access 2004–06
	2007	*2025 projected*	*2010*	*2025 projected*		
Lebanon	4.1	4.78	87%	89%	100%	98%
Lesotho	2.01	2.21	20%	28%	78%	36%
Liberia	3.75	6.77	62%	71%	64%	32%
Libya	6.16	8.09	86%	89%	–	97%
Lithuania	3.39	3.1	66%	69%	–	–
Luxembourg	0.47	0.57	82%	83%	100%	100%
Macedonia	2.04	2	72%	80%	100%	89%
Madagascar	19.68	29.95	28%	36%	47%	12%
Malawi	13.93	21.35	20%	28%	76%	60%
Malaysia	26.57	33.77	72%	80%	99%	94%
Maldives	0.31	0.41	32%	42%	83%	59%
Mali	12.34	20.59	33%	44%	60%	45%
Malta	0.41	0.43	97%	98%	100%	–
Marshall Islands	0.06	0.08	68%	73%	87%	82%
Mauritania	3.12	4.55	41%	48%	60%	24%
Mauritius	1.26	1.41	43%	49%	100%	94%
Mexico	106.54	124.7	77%	81%	95%	81%
Micronesia, Fed. Sts.	0.11	0.13	23%	27%	94%	25%
Moldova	3.79	3.5	48%	56%	92%	79%
Mongolia	2.63	3.11	58%	63%	72%	50%
Montenegro	0.6	0.61	–	–	98%	91%
Morocco	31.22	37.87	62%	70%	83%	72%
Mozambique	21.4	28.95	38%	50%	42%	31%
Namibia	2.07	2.56	38%	48%	93%	35%
Nauru	0.01	0.01	100%	100%	–	–
Nepal	28.2	38.86	18%	27%	89%	27%
Netherlands	16.42	16.96	83%	88%	100%	100%
New Zealand	4.18	4.76	87%	89%	–	–
Nicaragua	5.6	7.08	61%	67%	79%	48%
Niger	14.23	26.25	18%	24%	42%	7%
Nigeria	148.09	210.13	52%	63%	47%	30%
Niue	0	0	–	–	–	100%
Norway	4.7	5.23	78%	81%	100%	–
Oman	2.6	3.61	72%	75%	–	–
Pakistan	163.9	224.96	37%	46%	90%	58%
Palau	0.02	0.02	–	–	89%	67%
Palestine Authority	4.02	6.55	72%	76%	89%	80%
Panama	3.34	4.27	75%	82%	92%	74%
Papua New Guinea	6.33	8.57	14%	19%	40%	45%
Paraguay	6.13	8.03	62%	70%	77%	70%
Peru	27.9	34.15	74%	78%	84%	72%
Philippines	87.96	115.88	66%	75%	93%	78%
Poland	38.08	36.34	63%	68%	–	–
Portugal	10.62	10.71	61%	69%	99%	99%
Puerto Rico	3.99	4.33	99%	100%	–	–
Qatar	0.84	1.1	96%	97%	100%	100%
Romania	21.44	19.49	55%	60%	88%	72%
Russia	142.5	128.19	73%	74%	97%	87%

5 Total renewable water resources m³ per person per year 2003–07	6 Water dependency % of renewable water originating outside country 2007	7 Annual renewable surface water million m³ 2007 or latest	8 Annual renewable ground water million m³ 2007	9 Desalinated water produced million m³ per day 2007	
1,232	1%	3,803	3,200	47.3	Lebanon
1,684	0%	3,022	500	–	Lesotho
70,667	14%	232,000	45,000	–	Liberia
103	0%	200	500	18.0	Libya
7,257	38%	24,700	1,200	–	Lithuania
6,667	68%	3,100	80	–	Luxembourg
3,147	16%	6,400	0	–	Macedonia
18,112	0%	332,000	55,000	–	Madagascar
1,341	7%	17,280	2,500	–	Malawi
22,882	0%	566,000	64,000	4.3	Malaysia
91	0%	0	30	0.4	Maldives
7,398	40%	90,000	20,000	–	Mali
126	0%	1	50	31.0	Malta
–	–	0	0	0.7	Marshall Islands
3,715	96%	11,100	300	2.0	Mauritania
2,210	0%	2,358	893	–	Mauritius
4,272	11%	409,222	139,000	30.7	Mexico
–	-	0	0	–	Micronesia, Fed. Sts.
2,770	91%	11,650	400	–	Moldova
13,152	0%	32,700	6,100	–	Mongolia
–	–	0	0	–	Montenegro
921	0%	22,000	10,000	7.0	Morocco
10,970	54%	214,110	17,000	–	Mozambique
8,722	65%	15,655	2,100	0.3	Namibia
–	–	0	0	–	Nauru
7,747	6%	210,200	20,000	–	Nepal
5,583	88%	91,000	4,500	–	Netherlands
81,182	0%	0	0	–	New Zealand
35,847	4%	192,690	59,000	0.2	Nicaragua
2,411	90%	31,150	2,500	–	Niger
2,176	23%	279,200	87,000	3.0	Nigeria
–	-	0	0	–	Niue
82,684	0%	376,000	96,000	0.1	Norway
384	0%	1,050	955	109.0	Oman
1,410	76%	217,670	55,000	–	Pakistan
–	–	0	0	–	Palau
–	–	–	–	–	Palestine Authority
45,786	0%	144,670	21,000	–	Panama
136,063	0%	801,000	0	–	Papua New Guinea
54,563	72%	336,000	41,000	–	Paraguay
68,400	16%	1,913,000	303,000	5.4	Peru
5,767	0%	444,000	180,000	–	Philippines
1,599	13%	61,100	12,500	7.0	Poland
6,546	45%	68,700	4,000	1.6	Portugal
1,795	0%	0	0	–	Puerto Rico
65	4%	1	52	180.0	Qatar
9,761	80%	211,550	8,380	–	Romania
31,475	4%	4,231,250	788,000	–	Russia

Needs and Resources

	1 Total population millions		2 Urban population as % of total		3 Improved water source % of population with access 2004–06	4 Improved sanitation % of population with access 2004–06
	2007	*2025 projected*	*2010*	*2025 projected*		
Rwanda	9.73	15.22	24%	38%	65%	23%
St. Kitts and Nevis	0.05	0.06	32%	38%	99%	96%
St. Lucia	0.17	0.2	28%	33%	98%	89%
St. Vincent and Grenadines	0.12	0.13	48%	56%	–	–
Samoa	0.19	0.21	–	–	88%	100%
São Tomé and Principe	0.16	0.22	62%	72%	86%	24%
Saudi Arabia	24.74	34.8	82%	85%	–	–
Senegal	12.38	18	43%	50%	77%	28%
Serbia	9.86	9.96	53%	60%	99%	92%
Seychelles	0.09	0.09	55%	64%	88%	–
Sierra Leone	5.87	8.64	44%	56%	53%	11%
Singapore	4.44	5.1	100%	100%	100%	100%
Slovakia	5.39	5.31	57%	62%	100%	100%
Slovenia	2	1.94	52%	58%	–	–
Solomon Islands	0.5	0.71	19%	26%	70%	32%
Somalia	8.7	13.71	38%	46%	29%	23%
South Africa	48.58	52.3	62%	69%	93%	59%
Spain	44.28	46.62	77%	81%	100%	100%
Sri Lanka	19.3	20.33	15%	19%	82%	86%
Sudan	38.56	54.27	45%	57%	70%	35%
Suriname	0.46	0.48	76%	81%	92%	82%
Swaziland	1.14	1.24	26%	34%	60%	50%
Sweden	9.12	9.85	85%	87%	100%	100%
Switzerland	7.48	7.98	77%	82%	100%	100%
Syria	19.93	27.52	52%	58%	89%	92%
Tajikistan	6.74	8.93	24%	28%	67%	92%
Tanzania	40.45	59.99	26%	35%	55%	33%
Thailand	63.88	68.8	34%	42%	98%	96%
Togo	6.59	9.93	44%	55%	59%	12%
Tonga	0.1	0.11	–	–	100%	96%
Trinidad and Tobago	1.33	1.4	14%	21%	94%	92%
Tunisia	10.33	12.17	67%	73%	94%	85%
Turkey	74.88	89.56	70%	76%	97%	88%
Turkmenistan	4.97	6.07	48%	57%	72%	62%
Tuvalu	0.01	0.01	–	–	93%	89%
Uganda	30.88	54.01	13%	18%	64%	33%
Ukraine	46.21	39.88	69%	73%	97%	93%
United Arab Emirates	4.38	6.27	77%	79%	100%	97%
United Kingdom	60.77	65.19	90%	92%	100%	–
United States	305.83	354.93	82%	86%	99%	100%
Uruguay	3.34	3.55	93%	94%	100%	100%
Uzbekistan	27.37	33.96	37%	43%	88%	96%
Vanuatu	0.23	0.33	26%	34%	60%	50%
Venezuela	27.66	35.37	95%	97%	83%	68%
Vietnam	87.38	106.36	29%	38%	92%	65%
Yemen	22.39	36.57	29%	38%	66%	46%
Zambia	11.92	16.54	36%	42%	58%	52%
Zimbabwe	13.35	15.97	38%	47%	81%	46%

5 Total renewable water resources m³ per person per year 2003–07	6 Water dependency % of renewable water originating outside country 2007	7 Annual renewable surface water million m³ 2007 or latest	8 Annual renewable ground water million m³ 2007	9 Desalinated water produced million m³ per day 2007	
1,051	0%	9,500	7,000	–	Rwanda
558	0%	4	20	3.3	St. Kitts and Nevis
–	0%	0	0	–	St. Lucia
–	0%	0	0	–	St. Vincent and Grenadines
–	–	0	0	–	Samoa
13,885	0%	0	0	–	São Tomé and Principe
98	0%	2,200	2,200	1033.0	Saudi Arabia
3,328	34%	36,800	3,500	0.1	Senegal
19,851	79%	206,900	3,000	–	Serbia
–	0%	0	0	1.0	Seychelles
28,959	0%	150,000	25,000	–	Sierra Leone
139	0%	0	0	7.2	Singapore
9,276	75%	50,100	1,730	–	Slovakia
16,202	41%	31,720	13,500	–	Slovenia
93,515	0%	0	0	–	Solomon Islands
1,787	59%	14,400	3,300	–	Somalia
1,054	10%	48,200	4,800	18.0	South Africa
2,589	0%	109,800	29,900	100.2	Spain
2,410	0%	49,200	7,800	–	Sri Lanka
1,780	77%	62,500	7,000	0.4	Sudan
271,715	28%	122,000	80,000	–	Suriname
4,370	41%	4,510	660	–	Swaziland
19,246	2%	173,000	20,000	0.2	Sweden
7,377	24%	53,500	2,500	–	Switzerland
1,379	80%	22,710	5,550	–	Syria
2,456	17%	12,980	6,000	–	Tajikistan
2,512	13%	92,270	30,000	–	Tanzania
6,382	49%	398,735	41,900	–	Thailand
2,392	22%	14,000	5,700	–	Togo
–	–	0	0	–	Tonga
2,943	0%	0	0	36.0	Trinidad and Tobago
455	9%	3,400	1,595	13.0	Tunisia
2,918	2%	173,750	67,800	1.0	Turkey
5,115	97%	24,360	360	–	Turkmenistan
–	–	0	0	–	Tuvalu
2,290	41%	66,000	29,000	–	Uganda
3,002	62%	136,550	20,000	–	Ukraine
33	0%	150	120	950.0	United Arab Emirates
2,454	1%	146,200	9,800	33.3	United Kingdom
10,231	8%	2,913,000	1,300,000	580.0	United States
40,139	58%	139,000	23,000	–	Uruguay
1,896	77%	43,610	8,800	–	Uzbekistan
–	–	0	0	–	Vanuatu
46,102	41%	1,210,858	227,000	5.2	Venezuela
10,580	59%	878,210	48,000	–	Vietnam
195	0%	2,000	1,500	25.1	Yemen
9,016	24%	105,200	47,000	–	Zambia
1,537	39%	19,000	6,000	–	Zimbabwe

Water Uses

	Water use m³ per person 2000–05				Water use per sector as % of total 2000–05		
	1				**2**		
	Total	Domestic	Agriculture	Industry	Domestic	Agriculture	Industry
Afghanistan	898	16	881	0	2%	98%	0%
Albania	556	149	344	62	27%	62%	11%
Algeria	193	42	126	25	22%	65%	13%
Angola	24	5	14	4	23%	60%	17%
Antigua and Barbuda	77	46	15	15	60%	20%	20%
Argentina	772	130	572	73	17%	74%	9%
Armenia	967	289	636	43	30%	66%	4%
Australia	1,226	180	923	123	15%	75%	10%
Austria	260	91	2	166	35%	1%	64%
Azerbaijan	1,452	62	1,109	281	4%	76%	19%
Bahamas	–	–	–	–	–	–	–
Bahrain	492	245	219	28	50%	45%	6%
Bangladesh	592	19	570	4	3%	96%	1%
Barbados	299	112	75	149	38%	25%	50%
Belarus	281	66	85	131	23%	30%	47%
Belgium	831	101	10	779	11%	4%	85%
Belize	474	40	119	435	8%	25%	92%
Benin	17	5	8	4	32%	45%	23%
Bhutan	207	10	197	2	5%	95%	1%
Bolivia	160	21	134	11	13%	83%	7%
Bosnia and Herzegovina	–	–	–	–	–	–	–
Botswana	110	45	45	20	41%	41%	18%
Brazil	331	67	205	60	20%	62%	18%
Brunei	297	–	0	–	–	–	–
Bulgaria	1,331	41	250	1,041	3%	19%	78%
Burkina Faso	67	9	57	0	13%	86%	1%
Burma	679	8	667	4	1%	98%	1%
Burundi	42	7	33	2	17%	77%	6%
Cambodia	308	5	301	2	1%	98%	0%
Cameroon	64	12	47	5	18%	74%	8%
Canada	1,468	287	173	1,008	20%	12%	69%
Cape Verde	47	3	42	1	7%	91%	2%
Central African Republic	6	5	0	1	80%	4%	16%
Chad	26	5	22	–	17%	83%	0%
Chile	795	90	505	200	11%	64%	25%
China	485	32	329	125	7%	68%	26%
Colombia	246	124	113	9	50%	46%	4%
Comoros	14	7	6	1	48%	47%	5%
Congo	13	9	1	3	70%	9%	22%
Congo, Dem. Rep.	7	4	2	1	53%	31%	17%
Cook Islands	–	–	–	–	–	–	–
Costa Rica	654	193	349	112	29%	53%	17%
Côte d'Ivoire	54	13	35	6	24%	65%	12%
Croatia	–	–	–	–	–	–	–
Cuba	733	139	504	89	19%	69%	12%
Cyprus	297	84	217	4	28%	73%	1%
Czech Republic	251	102	6	143	41%	2%	57%
Denmark	236	76	100	59	32%	43%	25%

3 Water Footprint m³ per person water consumed directly or embedded 2001 or latest available	4 Irrigation Area equipped as % of total cultivated area 2008 or latest available	5 Hydropower As a % of total electricity produced 2005	6 Aquaculture Tonnes of fish, crustaceans and molluscs produced 2006	7 Wetlands of International Importance km² 2008	
660	40%	–	–	–	Afghanistan
1,228	49%	99%	1,970	831	Albania
1,216	7%	2%	288	29,596	Algeria
1,004	2%	66%	–	–	Angola
–	1%	–	–	36	Antigua and Barbuda
1,404	5%	32%	2,528	40,872	Argentina
898	51%	28%	1,056	4,922	Armenia
1,393	5%	6%	48,882	73,719	Australia
1,607	8%	57%	2,503	1,224	Austria
977	69%	14%	110	995	Azerbaijan
–	–	–	22	326	Bahamas
1,184	68%	0%	2	68	Bahrain
896	46%	6%	892,049	6,112	Bangladesh
1,355	32%	–	–	0	Barbados
1,271	2%	0%	4,150	2,831	Belarus
1,802	3%	0%	1,200	429	Belgium
1,646	3%	–	7,624	236	Belize
1,761	0%	1%	415	11,794	Benin
1,044	24%	–	–	–	Bhutan
1,206	4%	48%	430	65,181	Bolivia
–	0%	43%	7,621	109	Bosnia and Herzegovina
623	0%	0%	–	55,374	Botswana
1,381	4%	84%	271,696	64,341	Brazil
–	11%	0%	700	–	Brunei
1,395	3%	10%	3,257	203	Bulgaria
1,529	1%	–	200	2,992	Burkina Faso
1,591	15%	50%	574,990	3	Burma
1,062	2%	–	200	10	Burundi
1,766	7%	2%	34,200	546	Cambodia
1,093	0%	94%	340	6,066	Cameroon
2,049	2%	58%	170,938	130,667	Canada
995	6%	–	–	–	Cape Verde
1,083	0%	–	–	–	Central African Republic
1,979	1%	–	–	124,051	Chad
803	83%	48%	802,410	1,592	Chile
702	35%	16%	34,429,122	29,375	China
812	21%	77%	60,100	4,585	Colombia
–	0%	–	–	160	Comoros
–	0%	100%	21	4,390	Congo
734	0%	100%	2,970	–	Congo, Dem. Rep.
–	–	–	–	–	Cook Islands
1,150	20%	80%	19,962	5,101	Costa Rica
1,777	1%	26%	817	1,273	Côte d'Ivoire
–	0%	51%	14,897	866	Croatia
1,712	19%	1%	27,186	11,884	Cuba
2,208	28%	0%	2,667	16	Cyprus
1,572	1%	3%	20,431	547	Czech Republic
1,440	19%	0%	37,188	20,788	Denmark

Water Uses

	1 Water use m³ per person 2000–05				2 Water use per sector as % of total 2000–05		
	Total	Domestic	Agriculture	Industry	Domestic	Agriculture	Industry
Djibouti	25	21	4	–	84%	16%	0%
Dominica	210	–	0	–	–	–	–
Dominican Republic	398	128	263	7	32%	66%	2%
East Timor	–	–	–	–	–	–	–
Ecuador	1,340	167	1,102	71	12%	82%	5%
Egypt	977	76	844	57	8%	86%	6%
El Salvador	195	49	117	31	25%	60%	16%
Equatorial Guinea	230	191	2	36	83%	1%	16%
Eritrea	132	7	125	0	5%	95%	0%
Estonia	119	67	6	45	56%	5%	38%
Ethiopia	77	5	72	0	6%	94%	0%
Fiji	85	12	61	12	14%	71%	14%
Finland	476	65	13	398	14%	3%	83%
France	669	105	66	498	16%	10%	74%
Gabon	91	45	38	8	50%	42%	8%
Gambia	22	5	14	3	23%	65%	12%
Georgia	362	80	236	46	22%	65%	13%
Germany	570	70	113	387	12%	20%	68%
Ghana	47	11	31	5	24%	66%	10%
Greece	702	115	566	23	16%	81%	3%
Grenada	97	–	0	–	–	–	–
Guatemala	171	11	137	23	7%	81%	14%
Guinea	171	14	154	3	8%	90%	2%
Guinea-Bissau	121	16	99	6	13%	82%	5%
Guyana	2,195	40	2,142	13	2%	98%	1%
Haiti	120	6	114	1	5%	95%	1%
Honduras	128	10	102	15	8%	80%	12%
Hungary	751	70	241	440	9%	32%	59%
Iceland	523	174	1	348	33%	0%	67%
India	613	50	530	33	8%	86%	5%
Indonesia	386	31	352	3	8%	91%	1%
Iran	1,342	89	1,237	16	7%	92%	1%
Iraq	2,486	162	1,959	365	7%	79%	15%
Ireland	287	66	0	221	23%	0%	77%
Israel	291	106	168	17	36%	58%	6%
Italy	767	139	346	281	18%	45%	37%
Jamaica	157	54	77	27	34%	49%	17%
Japan	693	136	433	124	20%	62%	18%
Jordan	165	51	107	7	31%	65%	4%
Kazakhstan	2,352	40	1,923	388	2%	82%	17%
Kenya	80	14	63	3	17%	79%	4%
Kiribati	–	–	–	–	–	–	–
Korea, North	407	81	224	102	20%	55%	25%
Korea, South	393	140	189	65	36%	48%	16%
Kuwait	375	164	202	9	44%	54%	2%
Kyrgyzstan	1,983	63	1,859	61	3%	94%	3%
Laos	541	24	488	31	4%	90%	6%
Latvia	124	68	17	43	55%	14%	34%

3 Water Footprint m³ per person water consumed directly or embedded 2001 or latest available	4 Irrigation Area equipped as % of total cultivated area 2008 or latest available	5 Hydropower As a % of total electricity produced 2005	6 Aquaculture Tonnes of fish, crustaceans and molluscs produced 2006	7 Wetlands of International Importance km² 2008	
–	101%	–	–	30	Djibouti
–	–	–	–	–	Dominica
980	20%	15%	980	200	Dominican Republic
–	–	–	–	–	East Timor
1,218	29%	51%	78,300	2,011	Ecuador
1,097	100%	12%	595,030	1,057	Egypt
870	6%	35%	3,078	1,258	El Salvador
–	–	–	–	–	Equatorial Guinea
–	5%	–	–	–	Eritrea
–	1%	0%	703	2,260	Estonia
675	3%	99%	–	–	Ethiopia
1,245	1%	–	428	6	Fiji
1,727	3%	20%	12,891	7,995	Finland
1,875	14%	9%	238,860	8,662	France
1,420	1%	52%	126	17,638	Gabon
1,365	1%	–	–	312	Gambia
792	41%	86%	75	345	Georgia
1,545	4%	3%	35,379	8,712	Germany
1,293	0%	79%	1,150	1,784	Ghana
2,389	42%	8%	113,384	1,635	Greece
–	2%	–	–	–	Grenada
762	7%	43%	16,293	6,286	Guatemala
–	6%	–	–	64,224	Guinea
–	5%	–	–	391	Guinea-Bissau
2,113	30%	–	660	–	Guyana
848	8%	48%	–	–	Haiti
778	4%	32%	29,400	2,233	Honduras
789	3%	1%	14,686	2,354	Hungary
1,327	–	81%	8,241	590	Iceland
980	34%	14%	3,123,135	6,771	India
1,317	14%	8%	1,292,899	6,565	Indonesia
1,624	45%	9%	129,708	14,811	Iran
1,342	63%	2%	14,867	1,377	Iraq
–	0%	2%	53,122	670	Ireland
1,391	57%	0%	22,216	4	Israel
2,332	39%	11%	173,083	601	Italy
1,016	9%	2%	5,700	378	Jamaica
1,153	63%	7%	733,891	1,306	Japan
1,303	29%	1%	560	74	Jordan
1,774	13%	12%	528	3,533	Kazakhstan
714	2%	50%	1,012	1,018	Kenya
–	–	–	10	–	Kiribati
845	54%	57%	63,700	–	Korea, North
1,179	46%	1%	513,568	46	Korea, South
1,115	39%	0%	568	–	Kuwait
1,361	76%	87%	20	6,397	Kyrgyzstan
1,465	29%	–	78,000	–	Laos
684	0%	68%	565	1,484	Latvia

Water Uses

	1 Water use m³ per person 2000–05				2 Water use per sector as % of total 2000–05		
	Total	Domestic	Agriculture	Industry	Domestic	Agriculture	Industry
Lebanon	366	106	218	42	29%	60%	11%
Lesotho	28	11	6	11	40%	20%	40%
Liberia	34	9	19	6	27%	55%	18%
Libya	784	111	649	24	14%	83%	3%
Lithuania	78	61	6	12	78%	7%	15%
Luxembourg	108	110	0	711	–	–	–
Macedonia	–	–	–	–	–	–	–
Madagascar	873	24	835	13	3%	96%	2%
Malawi	84	12	67	4	15%	80%	5%
Malaysia	376	63	234	79	17%	62%	21%
Maldives	17	17	0	0	99%	0%	2%
Mali	530	48	477	5	9%	90%	1%
Malta	152	101	25	1	67%	17%	1%
Marshall Islands	–	–	–	–	–	–	–
Mauritania	606	53	534	18	9%	88%	3%
Mauritius	582	172	394	16	30%	68%	3%
Mexico	760	132	586	42	17%	77%	5%
Micronesia, Fed. Sts.	–	–	–	–	–	–	–
Moldova	544	52	179	313	10%	33%	58%
Mongolia	172	35	90	47	20%	52%	27%
Montenegro	–	–	–	–	–	–	–
Morocco	418	41	366	12	10%	87%	3%
Mozambique	34	4	29	1	11%	87%	2%
Namibia	153	37	109	7	24%	71%	5%
Nauru	–	–	–	–	–	–	–
Nepal	399	12	385	2	3%	96%	1%
Netherlands	494	30	167	296	6%	34%	60%
New Zealand	541	261	228	51	48%	42%	9%
Nicaragua	252	37	209	6	15%	83%	2%
Niger	173	7	165	1	4%	95%	0%
Nigeria	65	14	45	7	21%	69%	10%
Niue	–	–	–	–	–	–	–
Norway	481	110	51	321	23%	11%	67%
Oman	515	52	455	7	10%	88%	1%
Pakistan	1,138	22	1,093	23	2%	96%	2%
Palau	–	–	–	–	–	–	–
Palestine Authority	113	–	51	–	–	60%	–
Panama	268	180	75	13	67%	28%	5%
Papua New Guinea	14	7	0	5	50%	1%	38%
Paraguay	85	17	61	7	20%	71%	8%
Peru	752	63	614	76	8%	82%	10%
Philippines	362	60	268	34	17%	74%	9%
Poland	420	54	35	330	13%	8%	79%
Portugal	1,090	105	853	133	10%	78%	12%
Puerto Rico	–	–	–	–	–	–	–
Qatar	546	214	322	10	39%	59%	2%
Romania	1,056	91	602	363	9%	57%	34%
Russia	528	99	94	335	19%	18%	63%

3 Water Footprint m³ per person water consumed directly or embedded 2001 or latest available	4 Irrigation Area equipped as % of total cultivated area 2008 or latest available	5 Hydropower As a % of total electricity produced 2005	6 Aquaculture Tonnes of fish, crustaceans and molluscs produced 2006	7 Wetlands of International Importance km² 2008	
1,499	28%	10%	803	11	Lebanon
–	1%	–	2	4	Lesotho
1,382	0%	–	–	959	Liberia
2,056	22%	0%	480	1	Libya
1,128	0%	3%	2,224	505	Lithuania
–	0%	3%	–	3	Luxembourg
–	9%	21%	646	216	Macedonia
1,296	31%	–	11,213	7,876	Madagascar
1,277	2%	–	1,500	2,248	Malawi
2,344	5%	7%	168,317	1,342	Malaysia
–	–	–	–	–	Maldives
2,020	5%	–	1,000	41,195	Mali
1,916	30%	0%	1,126	0	Malta
–	–	–	–	690	Marshall Islands
1,386	9%	–	–	12,311	Mauritania
1,351	20%	–	443	4	Mauritius
1,441	23%	12%	158,642	59,526	Mexico
–	–	–	–	–	Micronesia, Fed. Sts.
1,474	14%	2%	4,470	947	Moldova
–	6%	–	–	14,395	Mongolia
–	–	–	11	200	Montenegro
1,531	16%	6%	1,161	2,720	Morocco
1,113	3%	100%	1,174	6,880	Mozambique
683	1%	97%	50	6,296	Namibia
–	–	–	–	–	Nauru
849	47%	100%	25,409	344	Nepal
1,223	43%	0%	43,945	8,189	Netherlands
–	8%	55%	107,522	391	New Zealand
819	3%	15%	11,220	4,057	Nicaragua
–	1%	–	40	43,176	Niger
1,979	1%	34%	84,578	10,767	Nigeria
–	–	–	–	–	Niue
1,467	14%	99%	708,780	1,164	Norway
1,606	56%	0%	89	–	Oman
1,218	80%	33%	121,825	13,436	Pakistan
–	–	–	5	5	Palau
–	–	–	–	–	Palestine Authority
979	5%	64%	8,744	1,599	Panama
2,005	0%	–	–	5,949	Papua New Guinea
1,165	2%	100%	2,100	7,860	Paraguay
777	28%	78%	28,393	67,840	Peru
1,543	15%	15%	623,369	684	Philippines
1,103	1%	1%	35,867	1,451	Poland
2,264	32%	10%	6,778	866	Portugal
–	–	–	266	–	Puerto Rico
1,087	62%	0%	36	–	Qatar
1,734	8%	34%	8,088	6,836	Romania
1,858	4%	18%	105,525	103,238	Russia

Water Uses

	1 Water use m³ per person 2000–05				2 Water use per sector as % of total 2000–05		
	Total	Domestic	Agriculture	Industry	Domestic	Agriculture	Industry
Rwanda	17	4	12	1	24%	68%	8%
Saint Kitts and Nevis	–	–	–	–	–	–	–
Saint Lucia	103	–	0	–	–	–	–
Saint Vincent and Grenadines	86	–	0	–	–	–	–
Samoa	–	–	–	–	–	–	–
São Tomé and Principe	53	–	0	–	–	–	–
Saudi Arabia	963	87	848	29	9%	88%	3%
Senegal	205	9	190	5	4%	93%	3%
Serbia	–	–	–	–	–	–	–
Seychelles	152	99	11	42	65%	7%	28%
Sierra Leone	78	4	72	2	5%	92%	3%
Singapore	82	37	3	42	45%	4%	51%
Slovakia	–	–	–	–	–	–	–
Slovenia	–	–	–	–	–	–	–
Solomon Islands	–	–	–	–	–	–	–
Somalia	401	2	399	0	0%	99%	0%
South Africa	268	84	168	16	31%	63%	6%
Spain	856	115	583	159	13%	68%	19%
Sri Lanka	623	15	594	15	2%	95%	2%
Sudan	1,091	29	1,054	8	3%	97%	1%
Suriname	1,519	68	1,406	45	4%	93%	3%
Swaziland	1,009	23	974	12	2%	97%	1%
Sweden	332	122	29	180	37%	9%	54%
Switzerland	356	86	7	264	24%	2%	74%
Syria	876	75	770	31	9%	88%	4%
Tajikistan	1,901	70	1,742	89	4%	92%	5%
Tanzania	143	15	128	1	10%	89%	0%
Thailand	1,391	35	1,322	34	2%	95%	2%
Togo	30	16	13	1	53%	45%	2%
Tonga	–	–	–	–	–	–	–
Trinidad and Tobago	240	162	15	62	68%	6%	26%
Tunisia	270	37	221	11	14%	82%	4%
Turkey	548	91	404	59	15%	75%	10%
Turkmenistan	5,322	91	5,192	41	2%	98%	1%
Tuvalu	–	–	–	–	–	–	–
Uganda	12	5	5	2	43%	40%	17%
Ukraine	781	95	410	276	12%	52%	35%
United Arab Emirates	889	137	737	15	15%	83%	2%
United Kingdom	161	35	5	121	22%	3%	75%
United States	1,654	210	682	761	13%	41%	46%
Uruguay	929	24	894	12	3%	96%	1%
Uzbekistan	2,292	109	2,136	47	5%	93%	2%
Vanuatu	–	–	–	–	–	–	–
Venezuela	330	150	157	23	46%	47%	7%
Vietnam	883	68	601	213	8%	68%	24%
Yemen	178	14	160	4	8%	90%	2%
Zambia	157	26	119	12	17%	76%	7%
Zimbabwe	329	46	260	23	14%	79%	7%

3 Water Footprint m³ per person water consumed directly or embedded 2001 or latest available	4 Irrigation Area equipped as % of total cultivated area 2008 or latest available	5 Hydropower As a % of total electricity produced 2005	6 Aquaculture Tonnes of fish, crustaceans and molluscs produced 2006	7 Wetlands of International Importance km² 2008	
1,107	1%	–	400	–	Rwanda
–	0%	–	–	–	Saint Kitts and Nevis
–	2%	–	–	1	Saint Lucia
–	–	–	–	–	Saint Vincent and Grenadines
–	–	–	–	–	Samoa
–	24%	–	–	0	São Tomé and Principe
1,263	46%	0%	15,586	–	Saudi Arabia
1,931	5%	10%	200	997	Senegal
–	1%	–	4,835	556	Serbia
–	4%	–	704	1	Seychelles
896	5%	–	–	2,950	Sierra Leone
–	–	0%	8,573	–	Singapore
–	13%	15%	1,263	407	Slovakia
–	2%	23%	1,369	82	Slovenia
–	–	–	–	–	Solomon Islands
671	15%	–	–	–	Somalia
931	10%	1%	3,352	5,440	South Africa
2,325	20%	7%	293,287	2,818	Spain
1,292	30%	39%	3,782	85	Sri Lanka
2,214	11%	30%	1,600	77,846	Sudan
1,234	76%	–	180	120	Suriname
1,225	26%	–	–	–	Swaziland
1,621	6%	46%	7,549	5,145	Sweden
1,682	6%	54%	1,214	87	Switzerland
1,827	25%	10%	8,902	100	Syria
939	68%	98%	26	946	Tajikistan
1,127	2%	59%	10	48,684	Tanzania
2,223	25%	4%	1,385,801	3,706	Thailand
1,277	0%	40%	3,020	12,104	Togo
–	–	–	5	–	Tonga
1,039	3%	0%	–	159	Trinidad and Tobago
1,597	8%	1%	2,775	7,265	Tunisia
1,615	19%	24%	129,073	1,795	Turkey
1,764	103%	0%	16	–	Turkmenistan
–	–	–	1	–	Tuvalu
–	0%	–	32,392	3,548	Uganda
1,316	8%	7%	4,030	7,447	Ukraine
–	89%	0%	570	–	United Arab Emirates
1,245	4%	1%	171,848	9,180	United Kingdom
2,483	12%	–	465,061	13,123	United States
–	13%	87%	37	4,249	Uruguay
979	87%	13%	3,800	5,584	Uzbekistan
–	–	–	114	–	Vanuatu
883	17%	74%	22,210	2,636	Venezuela
1,324	42%	40%	1,657,727	258	Vietnam
619	41%	0%	–	–	Yemen
754	3%	99%	5,125	40,305	Zambia
952	5%	57%	2,450	–	Zimbabwe

Glossary

access to an improved source – availability of an **improved water source** providing at least 20 litres per person per day within 1 kilometre of the dwelling.

aquifer – a natural underground layer, often of sand or gravel, that contains water; some aquifers are very deep and in hard rock, and have taken millions of years to accumulate their supply.

billion – a thousand million.

blue water – water withdrawn from lakes, rivers and aquifers for irrigation.

brackish water – water that is neither **fresh** nor **salt**.

desalination – the changing of **salt** or **brackish** water into **fresh** water; see also *thermal* **distillation** and **reverse osmosis**.

evaporation – the process of liquid water becoming water vapour, including vaporization from water surfaces, land surfaces, and snow fields, but not from leaf surfaces.

evapotranspiration – both **evaporation** and **transpiration** (the process by which water is evaporated from a plant surface, such as leaf pores).

fresh water – water that contains fewer than 1,000 milligrams per litre of dissolved solids.

green water – rain water, in the context of agriculture

groundwater – water that lies deep underground in aquifers. Normally free of contamination, it is regarded as a safe source of drinking water.

impoundment – control of water by dam or embankment to prevent water from flowing along its natural course.

improved sanitation – toilet facilities for the safe disposal of human excreta. These include "wet" systems, where water in a U-shaped pipe creates a seal, and which may be connected to a sewer or septic tank, and "dry" systems such as a pit toilet with a cleanable squat plate, cover and solidly constructed pit. Sanitation is considered adequate if it can effectively prevent human, animal and insect contact with faeces; this excludes public toilets in most settings.

improved water source – a water source with protection from contamination, such as a household connection to a safe piped supply or a public standpipe similarly connected; a borehole or protected well; a covered spring or rainwater collection system.

internal renewable water resources – average annual flow of rivers and recharge of groundwater generated from **precipitation** falling within the country's borders.

leaching – the process by which soluble materials in the soil, such as salts, nutrients, pesticide chemicals or contaminants, are washed into a lower layer of soil, or are dissolved and carried away by water.

melt-water – water produced by the melting of snow or ice.

precipitation – rain, snow, hail, sleet, dew, and frost.

renewable resources – total resources offered by the average annual natural inflow and run-off that feed a catchment area or aquifer; natural resources that, after exploitation, can return to their previous stock levels by the natural processes of growth or replenishment.

reverse osmosis – a desalination process that uses a semi-permeable membrane to separate and remove dissolved solids, viruses, bacteria and other matter from water; **salt** or **brackish** water is forced across a membrane, leaving the impurities behind and creating **fresh water**.

river basin – the area of land drained by a river and its tributaries. A basin is considered "closed" when its water is over-committed to human uses, and "closing" when it is approaching that state.

run-off – the movement of rain water over ground.

salt water – water that contains significant amounts of dissolved solids.

sewerage – a system of pipes with household connections to larger receptor and interceptor pipes and tunnels, that carries off waste matter, either to a treatment plant or directly to a river or stream.

surface water – water pumped from sources open to the atmosphere, such as rivers, lakes, and reservoirs.

unimproved water source – vendor, tanker trucks, and unprotected wells and springs.

thermal distillation – method of **desalination** in which water is boiled to steam and condensed in a separate reservoir, leaving behind contaminants with higher boiling points than water.

wastewater treatment – the process of turning contaminated water into water that can be re-used for a range of purposes, depending on the level to which it has been treated.

water table – the upper level of groundwater in soil.

withdrawal – water removed from groundwater or surface water for use.

Useful Conversions

1 cubic metre (m^3) = 1,000 litres

1 cubic kilometre (km^3) = 1,000,000,000 cubic metres (m^3) = 1,000,000,000,000 litres

1 litre = 0.264 US gallons (liquid) = 0.219 UK gallons

1 US gallon (liquid) = 3.785 litres

1 UK gallon = 4.55 litres

1 cubic metre (m^3) = 264.172 US gallons (liquid) = 219.9 UK gallons

1 US gallon (liquid) = 0.00378 cubic metres = 3,785 cubic centimetres (cc)

1 UK gallon = 0.00454 cubic metres = 4,546 cubic centimetres (cc)

1 cubic kilometre (km^3) = 810,713 acre feet

1 acre foot = 1,233 cubic metres (m^3) = 325,851 US gallons (liquid)

1 kilometre (km) = 0.621 miles

1 mile = 1.6 kilometres (km)

1 kilogram (kg) = 2.2 pounds (lb)

1 pound (lb) = 0.45 kilograms (kg) = 450 grams (g)

Metric water–weight conversion

1 kilogram (kg) of water = 1 litre of water

1 gram of water = 1 cubic centimetre (cc) of water

1 metric tonne (mt) of water = 1,000 kilogram (kg) of water = 1,000 litres of water = 1 cubic metre (m^3)

Sources

For sources that are available on the internet, in most cases only the root address has been given. To view the source, it is recommended that the reader types the title of the page or document into Google or another search engine.

20–21 The Global Water Pot
THE WORLD'S WATER
Shikolomanov I. World water resources at the beginning of the 21st century. http://webworld.unesco.org [accessed July 2008].
UNEP Vital Water Graphics. 5 and 15. www.unep.org/vitalwater
Water – a shared responsibility. The United Nations World Water Development Report 2. World Water Assessment Programme. Paris: UNESCO and New York: Berghahn Books. 2006. Chapter 4.

22–23 Water Shortage
WATER SHORTAGE
FAO Aquastat. www.fao.org [downloaded July 2008]
China statistical yearbook 2007. Beijing: China Press. 2008
SHORTAGES EXPECTED
United States General Accounting Office. GAO analysis of state water managers' responses to GAO survey. 2008. www.circleofblue.org
California
www.ens-newswire.com 2009 Feb 27
UNEQUAL DISTRIBUTION
CIA Factbook. www.cia.gov
FAO Aquastat www.fao.org
Australia
Australian Water Association. Australia water statistics. www.awa.asn.au
Brazil
Clarke R and King J. *O Atlas da Agua.* Sao Paolo, Publifolha. 2005. p.95.
By 2025 nearly 2 billion...
World Health Organization. 10 facts about water scarcity. 2008 June 4. www.who.int

24–25 Rising Demand
HOW WATER IS USED; WORLD WATER USE
FAO Aquastat. www.fao.org [downloaded July 2008]
INCREASING USE
Shikolomanov I. World water resources at the beginning of the 21st century. http://webworld.unesco.org [accessed July 2008].
1900: 350 cubic metres
Shiklomanov op.cit.

26–27 Dwindling Supply
FAO. The irrigation challenge. Issues paper 4. Rome: FAO, 2003. www.fao.org
Pearce F. Asian farmers sucking the continent dry. *New Scientist.* 2004 Aug 28.
Shar T. Groundwater: a global assessment of scale and significance. Chapter 10 in: *Comprehensive assessment of water management in agriculture.* London: Earthscan and Colombo: International Water Management Institute, 2007. p.399.
Zekster IS, Everett LG. Groundwater resources of the world and their use. IHP-VI, Series on Groundwater no. 6. UNESCO. 2004. www.unesco.org
GROUNDWATER
WHYMAP
Denver, Colorado, USA
Moore JE, Raynolds RG, Barkmann PE. Groundwater mining of bedrock aquifers in the Denver Basin – past present and future. *Environmental Geology.* 2004 Dec. 47:1 pp. 63-68. Berlin/Heidelberg: Springer.
Great Plains, USA
Brown L. Water tables falling and rivers running dry. 2007 July 24. Earth Policy Institute. www.earth-policy.org
New York Times map: Choking on growth. www.nytimes.com [accessed August 2008]
Gujarat, India
Black M, Talbot R. Water – a matter of life and health. Delhi: Unicef and OUP. 2005.
Brown L. op. cit.
Shah T, International Water Management Institute, cited in Depleting water level a big problem. *The Economic Times.* 2007 July 26 http://economictimes.indiatimes.com
Iran
Brown L. op .cit.
FAO Aquastat. www.fao.org [accessed August 2008]
Libya
Salem O. Water Resources Management in Libya. Workshop on Integrated Water Resources Management in Libya, 11-12 April 2007. p14. www.gwpmed.org
Mexico City, Mexico
Watkins K. Save a little water for tomorrow. *International Herald Tribune.* 2006 March 17. http://hdr.undp.org
Mexico–USA border
Hibbs B. Natural and induced transboundary ground water flows: Hueco Bolson Aquifer, El Paso/Juarez

area. 2007 May 1. http://ngwa.confex.com
What is the Paso del Norte. www.sharedwater.org
Navajo Aquifer, Arizona, USA
Grabiel T. Drawdown: an update on groundwater mining on Black Mesa. NRDC issue paper. 2006 March. www.nrdc.org
North China Plain
Yardley J. Beneath booming cities, China's future is drying up. *New York Times.* 2007 Sept 28. www.nytimes.com
Pearce F. op. cit.
Punjab, India and Pakistan
Brown L. op. cit.
Tianjin
Choking on growth. accompanying map. *New York Times.* www.nytimes.com [accessed August 2008]
Yemen
Sana'a University Water and Environment Centre www.wec.edu.ye/research.htm
Amel Al-Ariqi. Water war in Yemen. *Yemen Times.* 2006 March. 14:932. www.yementimes.org
FAO Aquastat. www.fao.org [accessed August 2008]
Brown L. op. cit.
1.5 billion...
800 cubic kilometres
Bundesanstalt fur Geowissenschaften under Rohstoffe (BGR). Groundwater. www.bgr.bund.de [accessed August 2008]

28–29 Competition and Conflict
WATER DEPENDENCY
FAO Aquastat www.fao.org
Bolivia–Chile
Newton J. The Silala river: a catalyst for cooperation? Abstract for paper at Sustainable Waters in a Changing World Conference. 2007 April 9. www.umass.edu
Central Asia
Linn JF. The impending water crisis in Central Asia. An immediate threat. The Brookings Institution. 2008 June 19 2008. www.brookings.edu
Linn JF. The compound water-energy-food crisis risks in Central Asia: update on an international response. The Brookings Institution. 2008 September 4. www.brookings.edu
Khamidov A. Ferghana valley: harsh winter's legacy stokes ethnic tension. *Eurasia Insight.* 2008 June 2. www.eurasianet.org
India
UNESCO. Sharing water. In: *The 2nd*

UN World Water Development Report: Water: a shared responsibility. Paris and New York: UNESCO and Berghahn Book. 2006. p.379.

Bangalore tense at river dispute. 2007 Feb 6. http://news.bbc.co.uk

Israel and Palestine

Human Development Report 2006/07. Beyond scarcity: power, poverty and the global water crisis. Chapter 6. UNDP. 2006. http://hdr.undp.org www.passia.org

The Nile Basin

Nile Basin www.internationalrivers.org

Around 260 river basins...

Jägerskog A and Phillips D. Managing trans-boundary waters for human development. Human Development Report Office. Occasional paper, 2006/9.

32–33 Climate Change

IPCC. *Climate change 2007.* Synthesis report. Adopted at IPCC Plenary XXVII, 2007 Nov 12-17.

Bates B et al. Climate change and water. IPCC Technical Paper VI, Executive Summary and sections 3.4.1.4 and 4.4.3. UNEP and WMO. 2008 June.

Human Development Report 2007/08. Fighting climate change: human solidarity in a divided world. Chapter 2. UNDP. 2007. http://hdr.undp.org.

CHANGE IN RUN-OFF

US Geological Survey. Published in: Bates B, Kundzewicz Z, Wu S, Palutikof J. Climate change and water. IPCC Technical Paper VI. June 2008. p.30.

Africa; Australia and New Zealand; Europe; South, South-East and East Asia
The collapse of the Asian "water tower"
IPCC Synthesis report. p.50.

Caribbean Islands; Central and West; Asia; Latin America; Pacific Islands

Human Development Report 2007/08, Chap 2, pp.95–98.

North America

Field CB et al. *Climate Change 2007: Impacts, Adaptation and Vulnerability. Contribution of Working Group II to the Fourth Assessment Report of the Intergovernmental Panel on Climate Change.* Parry ML et al editors. Cambridge: Cambridge University Press. 2007. p.627-9.

Schneider K. US faces era of water scarcity. 2008 July 9. www. circleofblue.org

Lake Mead, the American Southwest, and water: an interview with Tim Barnett. www.circleofblue.org

Alles D. The Colorado river. An ecological case study in coupled human and natural systems. Western Washington University. 2007 May. http://fire.biol.wwu.edu

POPULATION IN WATER-STRESSED RIVER BASINS

Alcamo et al. 2007. cited in Bates B et al. op. cit. p.45.

By 2080 up to 20%...

IPCC Synthesis report. p.49.

34–35 Urbanization
INCREASING URBAN POPULATIONS
UN-HABITAT 2005. Global Urban Observatory, Urban Indicators Programme, Phase III. Cited in: State of the world's cities report 2006/7.

UNEQUAL ACCESS

Hewett PC and Montgomery MR. Poverty and public services in developing-country cities. New York: Population Council. Cited in: Water and sanitation in world's cities, UN-HABITAT, 2003. p.66.

THE PRICE OF WATER

State of the world report 2007. Worldwatch. 2007. www.worldwatch.org

SEWAGE

Collignon B, Vezina M. *Independent water and sanitation providers in African cities.* Washington DC: World Bank. 2000. p.5. www.wsp.org

Urban population 2000...

World Urbanization Prospects: The 2007 Revision. http://esa.un.org/unup/ [downloaded July 2008]

36–37 Altered Flows

The new great walls. A guide to China's overseas dam industry. International Rivers 2008, July. www.internationalrivers.org

To dam or not to dam? Five years on from the World Commission on Dams. Dam right. WWF. 2005. www.panda.org

Dammed rivers, damned lies. International Rivers. 2008. www.internationalrivers.org

FRAGMENTED RIVERS

Map supplied by Landscape Ecology Group, Umeå University www.umu.se

Chalillo, Belize

To dam or not to dam? WWF, 2008.

Chinese dams

International Journal on Hydropower and Dams. cited in: *To dam or not to dam?* op. cit.
The new great walls. op. cit.

Indian dams

World's top ten rivers at risk. WWF, 2005. http://assets.panda.org/

Tipaimukh dam to kill the Meghna. *The New Nation.* 2008 Nov 2. http://nation.ittefaq.com

Patagonian rivers

Basic facts: Baker & Pascua rivers, proposed dams and transmission lines. International Rivers. www.internationalrivers.org

Mississippi Delta

Nilsson C, Reidy C. Dams in the world. In: *Dams under debate.* Formas. 2006. p6. www.formas.se

Murray-Darling, Australia

Murray-Darling 'never been worse' 2008 Sept 2. www.sbs.com.au

Sedimentation

FAO statistics.

WORLD DAMS

World Commission on Dams www.dams.org

RESERVOIR EMISSIONS

Warming the earth. In: *Dammed rivers, damned lies,* International Rivers 2008.

38–39 Draining Wetlands

Millennium Ecosystem Assessment. Ecosystems and human wellbeing: wetlands and water. Washington DC: World Resources Institute. 2005

PROTECTING WETLANDS

Ramsar Convention www.ramsar.org

Inner Niger Delta, Mali

Inner Niger Delta flooded savannah. Encyclopedia of Earth. 2008 September 2. www.eoearth.org

Tonle Sap, Cambodia

Millennium Ecosystem Assessment. op. cit. p.30.

Pearce F. Where have all the fish gone? The mighty Mekong is drying up - and so is the river's rich harvest. Vast new dams in China could be to blame. 2004 April 21. www.mongabay.com

Navarro P. Quake lakes spur rethinking of China's dam building strategy. *China Brief.* 8:12 June 2008. Jamestown Foundation. www.jamestown.org

DESTROYING THE GARDEN OF EDEN

UNEP project to help manage and restore the Iraqi marshlands http://imos.grid.unep.ch

Support for environmental management of the Iraqi marshlands http://marshlands.unep.or.jp

Bell JW. Ancient Sumeria. www.jameswbell.com

FLORIDA MANGROVES

Raines Ward D. Water wars: drought, flood, folly and the politics of thirst. Riverhead Books. New York. 2002.

Grunwald M. Sweet deal. *Time.* 2008 July 7. www.time.com

Williams WM. Florida deal for Everglades may help big sugar. *New York Times.* 2008 Sept 13. www.nytimes.com

Wetlands cover...

Millennium Ecosystem Assessment. op. cit.

The value of mangroves...

In the front line. Shoreline protection and other ecosystem services from mangroves and coral reefs. UNEP, 2006. pp.12-14. www.unep-wcmc.org

40–41 Drylands and Droughts
Australia
South Australia drought worsens 2008 July 10. http://news.bbc.co.uk

Drought Statement. A dry autumn and winter over most of Australia. 2008 Sept 3. www.bom.gov.au

Cyprus
Imported water saves Cyprus from drought. 2008 Aug 17. www.articlesbase.com

Doing battle with the Sahara
Berrahmouni N. Burgess N. Sahara desert. www.worldwildlife.org
EU and Africa to build 'green wall' across the Sahara. 2007 December 9. http://biopact.com

Gobi Desert – on the move
Desertification affects over 18% of Chinese territory. 2006 June 17. http://english.gov.cn
China suffers great losses from desertification. 2003 June 17. http://english.peopledaily.com.cn
Economy EC. The great leap backwards? 2007 Sept/Oct. www.foreignaffairs.org

Horn of Africa
Lederer EM. UN: Nearly 17 million need food in Horn of Africa. 2008 Sept 20. www.guardian.co.uk

Las Vegas
Moran T. Hinman K. Water wars: quenching Las Vegas' thirst. 2007 April 5. http://abcnews.go.com
Roberts C. Vegas heading for 'dry future'. 2005 July 29. http://news.bbc.co.uk
Tanner A. Las Vegas growth depends on dwindling water supply. 2007 Aug 21. www.reuters.com

Mongolia
Helminen M. Serious drought in Mongolia causes difficulties for herders. 2002 Sept 10. www.reliefweb.int

Spain
Lloyd Roberts S. Spain sweats amid 'water wars'. 2008 Aug 18. http://news.bbc.co.uk
Spain suffers worst drought. 2008 April 18 http://edition.cnn.com/

PEOPLE AFFECTED BY DROUGHT
EM-DAT: The OFDA/CRED International Disaster Database – www.emdat.net – Universite Catholique de Louvain, Brussels, Belgium.

The area of hyperarid land...
Climate change 2007. Synthesis report. Adopted at IPCC Plenary XXVII, 2007 Nov 12-17.
Executive Summary. p.3.

42–43 Floods
Plastic bags banned, blamed for west India floods. 2005 Aug 29. Reuters. www.planetark.com
Mississippi River towns rush against rising water. 2008 June 18. http://edition.cnn.com
Logan T. Why Bangladesh floods are so bad. 2004 July 27. http://bbc.co.uk

FINANCIAL COST; FLOODS; RISING WATERS
EM-DAT: The OFDA/CRED International Disaster Database – www.emdat.net – Universite Catholique de Louvain, Brussels, Belgium.

Bihar, India
International Rivers http://www.internationalrivers.org/

China
EM-DAT

Guatemala
Spark L. Painful legacy of Guatemalan storm. 2006 Oct 5. http://bbc.co.uk

46–47 Water for Drinking
DRINKING WATER; PROGRESS; WATER ACCESS; WATER COLLECTION; WATER TREATMENT
Progress on drinking water and sanitation. World Health Organization and United Nations Children's Fund Joint Monitoring Programme for Water Supply and Sanitation. New York: UNICEF and Geneva: WHO. 2008.
JMP Statistical database. www.wssinfo.org

48–49 Water for Sanitation
SANITATION; PROGRESS
Progress on drinking water and sanitation. World Health Organization and United Nations Children's Fund Joint Monitoring Programme for Water Supply and Sanitation. New York: UNICEF and Geneva: WHO. 2008. pp.42-50.

WASTEWATER TREATMENT
FAO Aquastat. www.fao.org [downloaded July 2008]

Nepal – wastewater solution
Thimi's community demonstrates how to manage waste water locally. 2006 Sept 1. www.unhabitat.org

2.5 billion people...
JMP op. cit. 2008

90% of sewage...
Sanitation: a wise investment for health, dignity, and development. Key messages for the International Year of Sanitation 2008 fact sheet. p. 6. UN economic and social development. http://esa.un.org

50–51 Water at Home
DOMESTIC WATER USE
FAO Aquastat. www.fao.org [downloaded Aug 2008]
Palestine figure calculated from data provided by www.passia.org

WATER IN THE HOME
Quickfacts. www.ec.gc.ca/water

The importance of hand washing
Child health week on course in Zambia. www.unicef.org

Hand-washing priorities
Graseff JA, Elder JP, Booth EM. Communication for health and behaviour change: a developing country perspective. 1993. cited in: Cairncross S, Yonli R. Domestic hygiene and diarrhoea – pinpointing the problem. Tropical Medicine and International Health. 5:1 pp.22-32, 2000 Jan. accessed from www.hygienecentral.org

Litres of water used daily...
Water Account Australia 2004-05 www.abs.gov.au
FAO Aquastat www.fao.org

Litres of water used per minute...
UK Environmental agency. Water resources, showers and baths. Table 1. www.environment-agency.gov.uk

52–53 Water and Disease
BURDEN OF DISEASE; DEATHS; WATER, SANITATION AND DISEASE
An estimated 10% ...
Prüss-Üstün A, Bos R, Gore F, Bartram J. Safer water, better health. Geneva: World Health Organization, 2008.

54–55 Disease Vectors
WHO disease information pages on neglected diseases, dracunculiasis, schistosomiasis. www.who.int

BLIGHTED LIVES; MALARIA DEATHS; YEARS OF DISABILITY
Global burden of disease 2004 update. Annex A Deaths and DALYS 2004: tables A1 and A2. www.who.int

WEST NILE VIRUS IN USA
Centers for Disease Control www.cdc.gov/ncidod/dvbid/westnile

A child dies...
Malaria. Fact sheet no. 94. www.who.int

56–57 Water for Food
Water for food, water for life: a comprehensive assessment of water management in agriculture. London: Earthscan, and Colombo: International Water Management Institute, 2007.

RISING DEMAND
Water for food, water for life. Table 3.1.

COMPARATIVE CALORIES
FAO potato: www.potato2008.org www.calorieking.com

CHANGING DIET IN CHINA
China statistical yearbook 2007. Beijing: China Press. 2008.

WATER FOR AGRICULTURE
FAO Aquastat. www.fao.org
Water for food, water for life. pp.100-01.

WATER FOR FOOD
Chapagain AK, Hoekstra AY. Water footprints of nations. vol 1: Main report. Nov 2004. Delft: UNESCO-IHE. 2004. p.41. www.waterfootprint.org

58–59 Dispossession by Water
THE DAMNED
Source for information on specific dams, unless otherwise stated: International Rivers www.internationalrivers.org

Bargi, Madhya Pradesh, India 1990
New Internationalist, 2001 July.

Chixoy, Guatemala
Johnston BR. Chixoy Dam Legacy Issues Study. Centre for Political Ecology, Santa Cruz, California, 2005 Mar. www.internationalrivers.org

Epupa, Namibia
Angola and Namibia plan huge dam. 2007 Oct 25. http://news.bbc.co.uk

Lesotho Highlands, Lesotho
Horta K and Pottinger L. A big idea for aiding Africa – think small *Los Angeles Times*. 2005 Sept 21. Accessed via www.internationalrivers.org

Manantali, Senegal
World Commission on Dams. *Dams and development. A new framework for decision making.* London: Earthscan. 2000, p.14. www.dams.org

Sobradinho, Bahia, Brazil
Movements occupy the hydroelectric plant of Sobradinho in the state of Barragens. 2008 June 10. www.mabnacional.org.br

Three Gorges, Yangtze
Yardley J. Chinese dam projects criticized for their human cost. *New York Times*. 2007 Nov 19. Accessed via http://internationalrivers.org

Tipaimukh, Manipur, India
Yumnam J. Damned hearings of proposed Tipaimukh Dam. www.kanglaonline.com

Yacyretá, Argentina-Paraguay
Final report of the Panel of the Independent Investigation Mechanism on Yacyretá Hydroelectric Project, 2004 Feb 27. Accessed via www.internationalrivers.org

By 2020, the Three Gorges...
China: four million more face relocation from the Three Gorges Reservoir area. 2007 Oct 11. www.redorbit.com

Dams and disease
International rivers www.internationalrivers.org

62–63 Irrigation
The 1st United Nations World Water Development Report: Water for people, water for life, chapter 8. UNESCO and Berghahn Book. 2003.
Water for food, water for life: a comprehensive assessment of water management in agriculture. London: Earthscan, and Colombo: International Water Management Institute, 2007.
Hamdy A. Water demand management in the Mediterranean. Paper prepared for 2nd Regional Workshop on Water Resource Management, held on 2–4 April 2000, at the Eastern Mediterranean University (EMU) in northern Cyprus. www.idrc.ca

IRRIGATED LAND
FAO Aquastat. www.fao.org [downloaded Oct 2008]

SPREAD OF IRRIGATION
Regional data available on www.fao.org [downloaded October 2008]

GROUNDWATER
Water for food, water for life. op. cit. Table 10.2.

UNEQUAL DEVELOPMENT
World Water Assessment Programme facts: www.unesco.org [accessed Dec 2008]

64–65 Water for Industry
INDUSTRIAL WATER USE
FAO Aquaastat. www.fao.org [downloaded Oct 2008]

INCREASING INDUSTRIAL USE; MAKING WATER WORK
Water used for industrial purposes...
World Water Assessment Programme facts: www.unesco.org [accessed Oct 2008]

WATER FOR BIO-FUELS
Water Footprint Newsletter. 2008 Nov. www.waterfootprint.org

66–67 Water for Energy
Shiklomanov KA. State Hydrological Institute (SHI), St Petersburg and United Nations Educational, Scientific and Cultural Organisation (UNESCO), 1999. www.grida.no
World Commission on Dams. *Dams and development. A new framework for decision making.* London: Earthscan. 2000, p.14. www.dams.org
United Nations Industrial Development Organization (UNIDO). Water and energy. In: *Water. A shared responsibility.* Paris: World Water Assessment Programme and New York: Berghahn Books, 2006. p.322.
Martinot E. Renewables 2007. Global status report. Table R4. *Renewable Energy Policy Network for the 21st Century.* www.ren21.net
World Water Assessment Programme facts: www.unesco.org

HYDROPOWER
Hydropower provides 19%...
World Development Indicators online. [downloaded Nov 2008]
China statistical yearbook 2007. Beijing: China Press. 2008.

LARGE AND SMALL HYDROPOWER
47% of China's hydroelectricity...
Martinot E. op. cit.

WATER LOSS IN USA
Torcellini P, Long N, Judkoff R. Consumptive water use for US power production. National Renewable Energy Laboratory. US Department of Energy, 2003 Dec. www.nrel.gov

68–69 Water for Fisheries
Bestari N et al. Case Study 1: Overview of small-scale freshwater aquaculture in Bangladesh. In: *Special evaluation study on small-scale freshwater rural aquaculture development for poverty reduction. Bangladesh, Philippines and Thailand.* Livelihoods Connect. Institute of Development Studies,

2008. www.livelihoods.org
New Internationalist. 325. 2000 July. quoting FAO statistics.
World Water Vision, World Water Council 2000. p 16.
Water. A shared responsibility. Paris: World Water Assessment Programme and New York: Berghahn Books, 2006. p 248.
Halweil B. Farming fish for the future. 2008 Oct. www.worldwatch.org

AQUACULTURE; PRODUCTION SHARE; INLAND WATERS; Aquaculture production increased...
Yearbooks of Fishery Statistics ftp.fao.org/fi/stat/summary/default.htm

Chile
International Aquaculture search. http://aquaculturenews.aquafeed.co.uk
Table A-4. World aquaculture production of fish, crustaceans, molluscs, etc. by principal producers in 2006. www.fao.org

China
China statistical yearbook 2008. Beijing: China Press. 2009. Table 12-20.

West and Central Africa
Inland waters contributed...
Dugan P et al. Chapter 12. Inland fisheries and aquaculture. In: *Water for food, water for life: a comprehensive assessment of water management in agriculture.* London: Earthscan, and Colombo: International Water Management Institute, 2007.

70–71 Transport and Leisure
WATERBORNE FREIGHT
Mississippi
Preliminary waterborne commerce statistics for 2007. Institute for Water Resources, Navigation Data Center. US Army Corps of Engineers. 2008 Nov 5.

INLAND WATERWAYS
Great Lakes
Waterborne commerce of the United States 2006 – part 3 waterways and harbors, Great Lakes. Compiled under the supervision of the Institute for Water Resources US Army Corps of Engineers, Alexandria, Virginia. US Army Corps. Navigation Data Center. Waterborne Commerce Statistics Center. IWR-WCUS-06-3 www.iwr.usace.army.mil

China's waterways
China statistical yearbook 2008. Beijing: China Press. 2009. Tables 15-4 and 15-8.

Golf courses
Golf courses in dry climates...
Golf course water efficiency introduction. www.allianceforwaterefficiency.org
Golf course water use. http://hillcountrywater.org

72–73 Water for Sale
PRICE OF PIPED WATER
World Water Assessment Programme
facts: www.unesco.org
RELATIVE COST OF WATER
Water and sanitation in the world's
cities. UN-Habitat, Nairobi. London:
Earthscan. 2003.
BOTTLED WATER
Hailes J. Bottled water – eau no! *Daily
Telegraph*. 2008 Nov 10.
www.telegraph.co.uk
Gies E. Rising sales of bottled water
trigger strong reaction from US
conservationists. *International Herald
Tribune*. 2008 March 19.
www.iht.com
Bottled Water. Pacific Institute
www.pacificinstitute.org
Gleick PH and Morrison J. Water risks
that face business and industry. In:
Gleick PH et al. *The world's water
2006–2007*. Island Press. p.165
Gleick PH. Bottled water: an update. In:
Gleick PH et al. op. cit. p169.
Zenith figures from: http://www.
britishbottledwater.org/vitalstats2.html
War on Want www.waronwant.org
India Resource Centre:
www.indiaresource.org
Production of plastic...
Pacific Institute Bottled water Fact Sheet
Dec 2007. www.pacificinstitute.org
Fewer than 20%...
Goodbye to bottles for Chicago Botanic
Gardens. 2008 June 13. www.bgci.org
**INCREASING GLOBAL
CONSUMPTION**
Gleick PH and Morrison J. op. cit.
p.165.
Gies E. op .cit.
TOP CONSUMERS
Gleick PH. Bottled water: an update. In:
Gleick PH et al. op.cit. p.169.

76–77 Water Pollutants
World Water Assessment Programme
facts: www.unesco.org
Water: a matter of life and death.
www.un.org/events
Persistent organic pollutants. A global
issue, a global response. www.epa.gov
**ORGANIC AND NON-ORGANIC
POLLUTANTS**
Water – a shared responsibility.
The United Nations World Water
Development Report 2. World
Water Assessment Programme. Paris:
UNESCO and New York: Berghahn
Books. 2006. Table 4.5. p.141.
HAZARDOUS WASTE
Water facts www.unesco.org
ORGANIC POLLUTION
World Development Indicators online.
[accessed Nov 2008]
EU POLLUTION
EEA – The European Pollutant Emission
Register (EPER).
http://eper.eea.europa.eu

Developing countries are...
World Water Assessment Programme
facts: www.unesco.org

78–79 Water Pollution
Water – a shared responsibility.
The United Nations World Water
Development Report 2. World
Water Assessment Programme. Paris:
UNESCO and New York: Berghahn
Books. 2006. Chapters 4 and 8.
Larsen J. Earth Policy Institute. Dead
zones increasing in world's coastal
waters. www.grinningplanet.com
Study shows continued spread of "dead
zones" e! Science News. 2008 Aug 14
http://esciencenews.com
Science focus: dead zones. Nasa.
http://disc.gsfc.nasa.gov
The number of dead zones...
Study shows continued spread of "dead
zones" op. cit.
FLUORIDE
WHO fact sheet. www.who.int
Black M, Talbot R. *Water, a matter of life
and health*, OUP New Delhi, 2005,
p.5.
CHRONIC POLLUTION
Baltic Sea
Do nitrogen cuts benefit the Baltic Sea?
ACS publications. 2006, May 17.
http://pubs.acs.org
Larsen J. op. cit.
China
Economy EC. The great leap backward?
Council on Foreign Relations. Foreign
Affairs, 2007 Sept/Oct.
www.foreignaffairs.org
Agence France-Presse citing *South
China Morning Post*, A6. 2005 Dec 29
Gulf of Mexico
Faeth P, Mehan T. Nutrient runoff
creates dead zone. World Resources
Institute, 2005 Jan.
http://archive.wri.org
India – human waste
Centre for Science and Environment.
Sewage canal: How to clean the
Yamuna. slide 18. www.cseindia.org
Indian Environmental Portal. Water
Pollution.
http://indiaenvironmentportal.org.in
Black M, Talbot R. op. cit.
India – pesticide pollution
Centre for Science and Environment.
CSE releases new study on pesticides
in soft drinks. 2007 March 11.
www.cseindia.org

80–81 Damaged Waterways
Water – a shared responsibility.
The United Nations World Water
Development Report 2. World
Water Assessment Programme. Paris:
UNESCO and New York: Berghahn
Books. 2006. Chapters 4 and 8.
POLLUTION OF THE HUANG HE
Larmer B. Bitter waters. In: China: inside
the dragon. *National Geographic*

special issue, 2008, May. pp. 146-69
http://ngm.nationalgeographic.com
POLLUTION INCIDENTS
Baia Mare tailings dams, Romania
Water – a shared responsibility. Chapter
8, p.283.
Bacsujlaky M. Examples of modern
mines that damaged rivers & fishers.
2004 Oct.
www.bristolbayalliance.com
Earthquake, Turkey
Water – a shared responsibility. op. cit.
p.281.
Oil slicks on Mississippi
Oil spill idles 200 ships in Mississippi
River. 2008 July 25. Associated Press.
www.msnbc.msn.com.
Nossiter A. Mississippi River reopened
after oil spill. 2008 July 25.
www.nytimes.com
Songhua River
Zhang X. The water pollution incident: a
test for China.
www.chinatoday.com.cn
Harbin water supply to resume as
schedule. 2005 Nov 26. www.gov.cn
Yamuna River, India
Pepper D. India's rivers are drowning in
pollution. Fortune. 2007 June 4.
CSE. Sewage canal. How to clean the
Yamuna. CSE, 2007. www.cseindia.org
We can bring the Yamuna back to life.
2007 April 18. CSE press release.

82–83 Threatened Ecologies
Finlayson CM, D'Cruz R et al. Inland
water systems. Chapter 20 of
Ecosystems and human well-being:
wetlands and water. Millennium
Ecosystem Assessment. Washington
DC: World Resources Institute. 2005.
Finlayson CM, D'Cruz R, Davidson N.
Ecosystems and human well-being:
wetlands and water. Synthesis report.
Millennium Ecosystem Assessment.
Washington DC: World Resources
Institute.2005
Freshwater Ecosystems of the World
– WWF/TNC. 2008. www.feow.org
Cosgrove WJ, Rijsberman FR. World
water vision. World Water Council,
2000, p 16.
Populations of freshwater species...
Finlayson CM and D'Cruz R et al.
Inland water systems. op. cit. p.562.
ENDEMIC MOLLUSCS
Finlayson CM and D'Cruz R et al. op.
cit. p.563.
**RELATIVE SPECIES RICHNESS
ENDANGERED SPECIES**
Finlayson CM, D'Cruz R, Davidson N.
Synthesis report. op. cit. p.26.
**Waterbirds and wetland-dependent
birds**
Stroud DA et al, compilers. Status of
migratory wader populations in Africa
and Western Eurasia in the 1990s.
International Wader Studies 2004: 15.
pp.1–259. cited in: Water population

estimates, fourth edition. Wageningen, The Netherlands Wetlands International, 2007.

Aquatic amphibians
Global Amphibian Assessment. www.globalamphibians.org

86–87 Millennium Development Goals
Progress on drinking water and sanitation. JMP. 2008. www.who.int
Safer Water, Better Health. Geneva: WHO. 2008 www.who.int
Millennium Development Goal website www.un.org/millenniumgoals
ON TRACK OR NOT; PROGRESS
Progress on drinking water and sanitation. op. cit.

88–89 Treaties and Obligations
WATER TREATIES
Transboundary Freshwater Dispute Database, Department of Geosciences, Oregon State University, 2008. www.transboundarywaters.orst.edu
BASIN TREATIES
Product of the Transboundary Freshwater Dispute Database, Department of Geosciences, Oregon State University. Additional information about the TFDD can be found at: www.transboundarywaters.orst.edu
A HUMAN RIGHT
UN Human Rights Council moves forward on the right to water and sanitation www.righttowater.org.uk

90–91 Deepening Co-operation
THE NEED FOR CO-OPERATION
Lake Chad
Odada EO. Lake Chad. Experience and lessons learned brief. 2006 Feb 7. www.ilec.org.jp
The Mekong
International Rivers www.internationalrivers.org
Mekong River Commission www.mrcmekong.org
Human Development Report 2006/07. Beyond scarcity: power, poverty and the global water crisis. Chapter 6. UNDP. 2006. http://hdr.undp.org
The Nile
Nile Basin Initiative www.nilebasin.org
The Colorado
Arizona Department of Water Resources www.water.az.gov
Living Rivers www.livingrivers.org
The Danube
Human Development Report op. cit.
International Commission for the Protection of the Danube River. www.icpdr.org
The Ganges and Brahmaputra
WATER ISSUES; CONFLICT AND CO-OPERATION
Transboundary Freshwater Dispute Database, Oregon State University, 2008.

www.transboundarywaters.orst.edu

92–93 Managing Water
CORRUPTION IN WATER; MANAGEMENT
Corruption adds...
Global corruption report 2008: corruption in the water sector. Transparency International and Cambridge University Press, 2008.
INTEGRATED WATER RESOURCES; MANAGEMENT
Ferghana Valley
Uganda
Status report on integrated water resources management and water efficiency plans. Prepared for the 16th session of the Commission on Sustainable Development. UN-Water, 2008.
Local water management
Harnath Jagawat, Revival through rivers. In: Making water everybody's business: practice and policy of water harvesting. Agarwal A, Narain S, Khurana I, editors, Delhi: CSE. 2001, p 105.
Down to Earth. Delhi: CSE. 2000 Jan 15.
Women and water management
WaterAid. www.wateraid.org

94–95 Water Footprint
Hoekstra AY, Chapagain AK. Water footprints of nations: Water use by people as a function of their consumption pattern. *Water Resource Management.* 2007. 21:35-48.
EMBEDDED WATER
waterfootprint.org
CONTRASTING WATER PROFILES
16% of water...
Hoeskstra AY, 2007. op. cit.
WATER USE
FAO Aquastat. www.fao.org

96–97 Water at a Price
AS SUPPLIES DIMINISH, COSTS INCREASE
Tortajada C, Castelan E. Water management for a megacity: Mexico City Metropolitan area. *Ambio*, 2003 March. 32:2 http://ambio.allenpress.com
Irrigation resources and water costs. www.ers.usda.gov
WATER PRIVATIZATION
Bolivia
New Internationalist. 338. 2001 Sept.
France
www.suez-environnement.com
Plus archive reports.
Godoy J. Is the water privatization trend ending? IPS News. 2008 June 30. www.alternet.org
Guyana
UK wastes millions of pounds of aid on failed water privatisation in Guyana. World Development Movement. 2007 Feb 19. www.waterjustice.org

Tanzania
UK water company fails in $20 million compensation claim from Tanzanian government. World Development Movement 2008 28 July. www.wdm.org.uk
Turkey
Hoedeman O, Senalp O. Turkey's government plans sweeping water privatization in run-up to World Water Forum in Istanbul, April 2008. www.corporateeurope.org

98–99 Technological Fixes
Gleick PH, Cooley H, Wolff G. With a grain of salt. In: *The world's water 2006/07.* Washington DC: Island Press. 2006. pp.51-89.
Tapping the oceans. *The Economist.* 2008 June 5. www.economist.com
DESALINATION
FAO Aquastat. www.fao.org [downloaded Jan 2009]
Spain
USA
Tapping the oceans. op. cit.
London
Thames water desalination plant, London, United Kingdom. www.water-technology.net
There are over 13,000...
Gleick PH. 2006. op. cit.
WATER FOR FOOD
Seawater Greenhouse Ltd. www.seawatergreenhouse.com
Check-dams
India Water Portal flikr photo collection. Shree Patel collection: Sisandra, Madaka, Kerala.

102–09 Needs and Resources
Col 1: UN Population Division. World population prospects: the 2006 revision. Table A.2.
Col 2: UN Population Division. World urbanization prospects: the 2005 revision. Table A.2.
Cols 3 & 4: UNICEF and WHO Joint Monitoring Programme.
Cols 5 – 9: FAO Aquastat, downloaded Oct 2008–Jan 2009.

110–17 Water Uses
Cols 1, 2 & 4: FAO Aquastat.
Col 3: www.waterfootprint.org
Col 5: World Development Indicators online. Downloaded Nov 2008.
Col 6: FAO. Yearbook of Fishery Statistics 2008.
Col 7: www.ramsar.org

Index